Research Methods in Plant Science

Research Methods in Plant Science

Vol. 2

Dr. Chandra P. Singh

AGROTECH PRESS
JAIPUR ▪ NEW DELHI

AGROTECH PRESS

Regd. Off. "Bony Residency", Gate No. 2
Opp. Tilak Public School, Vishweswariya Nagar,
Gopalpura Road, Jaipur - 302018, India
Ph: 0141 - 2761280, Telefax: 2761381
E-mail : agrotechpress@gmail.com

Delhi Office: 102, 1st Floor, Satyam House,
4327/3, Ansari Road, Darya Ganj,
New Delhi - 110002, India
Ph: 011-45652440
visit us: www.agrotechpress.com

Distributed By
KEYA INTERNATIONAL PRESS LLP
E-mail: keyainternationalpressllp@gmail.com

Published in India, 2025

ISBN : 9789384568542 (2 Vols. Set)

Typeset by Thomson Press

Printed and bound in India by Thomson Press Ltd.,
New Delhi

Contents

Preface

Agroforestry is any sustainable land-use system that maintains or increases total yields by combining food crops (annuals) with tree crops (perennials) and/or livestock on the same unit of land, either alternately or at the same time, using management practices that suit the social and cultural characteristics of the local people and the economic and ecological conditions of the area. It combines agricultural and forestry technologies to create more diverse, productive, profitable, healthy, and sustainable land-use systems. Agroforestry systems can be advantageous over conventional agricultural, and forest production methods. They can offer increased productivity, economic benefits, and more diversity in the ecological goods and services provided. Agroforestry provides a different land use option, compared with traditional arable and forestry systems. It makes use of the complementarity between trees and crops, so that the available resources can be more effectively exploited. It is a practice that respects the environment and has an obvious landscape benefit. Efficient, modern versions of agroforestry have been developed, that are adapted to the constraints imposed by mechanisation. The agroforestry plot remains productive for the farmer and generates continuous revenue, which is not the case when arable land is exclusively reforested. Agroforestry allows for the diversification of farm activity and makes better use of environmental resources.

Agroforestry is practiced for a variety of objectives. It represents an interface between agriculture and forestry and encompasses mixed land-use practices. These practices have been developed primarily in response to the special needs and conditions of tropical developing countries that have not been satisfactorily addressed by advances in conventional agriculture or forestry. The term is used to denote practices ranging from simple forms of shifting cultivation to complex hedgerow intercropping systems; systems including varying densities of tree stands ranging from widely-scattered Faidherbia (Acacia) albida trees in Sahelian millet fields, to the high-density multistoried homegardens of the humid tropics; and systems in which trees play a predominantly service role to those in which they provide the main commercial product. Detailed descriptions of a variety of such systems in the tropics are now available. It needs to be reemphasized that one concept is common to all these diverse agroforestry systems: the purposeful growing or deliberate retention of trees with crops and/or animals in interacting combinations for multiple products or benefits from the same management unit. In short, agroforestry has an important role to play both for food and wood security and the conservation of the environment. By integrating tree growing with crop production, the problems of poor agricultural production, worsening wood shortages and environmental degradation can be addressed. Furthermore, Agroforestry technologies/ practices are seen as an opportunity to take pressure off the remaining natural forests and to increase the diversity of vegetation on existing farms.

Biodiversity in agroforestry systems is typically higher than in conventional agricultural systems. With two or more interacting plant species in a given land area, it creates a more complex habitat that can support a wider variety of birds, insects, and other animals. Agroforestry is relevant to almost all environments and is a potential response to common problems around the globe, and agroforestry systems can be advantageous compared to conventional agriculture or forestry.

The concept of agro-forestry has been advocated due to environmental problems because there is serious concern over degradation of environment. The serious adverse ecological manifestation such as increase of carbon dioxide, in the atmosphere, global warming soil losses, repeated droughts, floods and pollution etc. are the result of dwindling forest resources. An agro-forestry system will help to increase the tree cover. It also makes available to people the required quantity of timber, fruit, fuel -wood, fodder etc. for which they traditionally depend on forests. Thus this system helps in reducing the pressure on forests and helps in conservation and development. The best safeguard against landslides can be through agro-forestry but this requires careful selection of tree species, grasses etc. The agro-forestry also maintains soil fertility through re-cycling of nutrients and prevents soil erosion and loss nutrients through leaching and run off. Many leguminous tree species fix nitrogen from the atmosphere and return much more in leaf fall than they take from the soil.

This book lays down schemes for an urgent, strong and sustained effort focused on forest management and governance, given the crucial role of forests in confronting some of the most important challenges of our time: climate change, poverty, and food security. This book covers various aspects including research, technology development, tools, training, and outreach activities which will be useful for the advancement of agroforestry. This will be an essential guide for agro-scientists, agriculture professionals and natural resource professionals who work with farmers, ranchers, woodland owners, tribes, and communities.

Dr. Chandra P. Singh

1

An Introduction to Forestry Agroforestry

Forestry is the science, art, and craft of creating, managing, using, conserving, and repairing forests and associated resources to meet desired goals, needs, and values for human benefit. Forestry is practised in plantations and natural stands. The main goal of forestry is to create and implement systems that manage forests to provide environmental supplies and services. The challenge of forestry is to create systems that are socially accepted while sustaining the resource and any other resources that might be affected. Silviculture is a process for creating, maintaining, or restoring an appropriate balance of essential components, structures, and functions that ensure long-term ecosystem vitality, stability and resiliency (Nyland, 2007). This is done at the ground level which can contain many varieties of trees. Modern forestry generally embraces a broad range of concerns, including ecosystem services by assisting forests to provide timber as raw material for wood products, wildlife habitat, natural water quality management, recreation, landscape and community protection, employment, aesthetically appealing landscapes, biodiversity management, watershed management, erosion control, and preserving forests as 'sinks' for atmospheric carbon dioxide. A practitioner of forestry is known as a forester.

Forest ecosystems have come to be seen as the most important component of the biosphere, and forestry has emerged as a vital applied science, craft, and technology.

History

In the 5th century, monks in the then Byzantine Romagna on the Adriatic coast, established a plantation of stone pine to provide fuelwood and food. This was the beginning of the massive forest mentioned by Dante Alighieri in his 1308 poem Divine Comedy.

Formal forestry practices were developed by the Visigoths in the 7th century when, faced with the ever increasing shortage of wood, they instituted a code concerned with the preservation of oak and pine forests. The use and management of many forest resources has a long history in China, dating from the Han Dynasty and taking place under the landowning gentry. It was also later written of by the Ming Dynasty Chinese scholar Xu Guangqi (1562–1633).

In Europe, control of the land included hunting rights, and though peasants in many places were permitted to gather firewood and building timber and to graze animals, hunting rights were retained by the members of the nobility. Systematic management of forests for a sustainable yield of timber is said to have begun in the German states in the 14th century, e.g. in Nuremberg, and in 16th-century Japan. Typically, a forest was divided into specific sections and mapped; the harvest of timber was planned with an eye to regeneration.

The practice of establishing tree plantations in the British Isles was promoted by John Evelyn, though it had already acquired some popularity. Louis XIV's minister Jean-Baptiste Colbert's oak Forest of Tronçais, planted for the future use of the French Navy, matured as expected in the mid-19th century: "Colbert had thought of everything except the steamship,"

Fernand Braudel observed. Schools of forestry were established beginning in the late 18th century in Hesse, Russia, Austria-Hungary, Sweden, France and elsewhere in Europe. During the late 19th and early 20th centuries, forest preservation programmes were established in British India, the United States, and Europe. Many foresters were either from continental Europe (like Sir Dietrich Brandis), or educated there (like Gifford Pinchot).

The enactment and evolution of forest laws and binding regulations occurred in most Western nations in the 20th century in response to growing conservation concerns and the increasing technological capacity of logging companies.

Tropical forestry is a separate branch of forestry which deals mainly with equatorial forests that yield woods such as teak and mahogany. Sir Dietrich Brandis is considered the father of tropical forestry.

Early Journals which are Still Present

Sylwan first published in 1820

Schweizerische Zeitschrift für Forstwesen first published in 1850.

The Indian Forester first published in 1875.

Šumarski list (Forestry Review, Croatia) was published in 1877 by Croatian Forestry Society.

Montes (Forestry, Spain) first published in 1877.

Revista pădurilor (Journal of Forests, Romania, 1881–1882; 1886–present), the oldest extant magazine in Romania

Forestry Quarterly, first published in 1902 by the New York State College of Forestry.

Šumarstvo (Forestry, Serbia) first published in 1948 by the Ministry of Forestry of Democratic Federal Yugoslavia, and since 1951 by Organ of Society of Forestry Engineers and Technicians of the Republic of Serbia (succeeding the former Šumarski glasnik published from 1907-1921)

As a science

Over the past centuries, forestry was regarded as a separate science. With the rise of ecology and environmental science, there has been a reordering in the applied sciences. In line with this view, forestry is one of three primary land-use sciences. The other two are agriculture and agroforestry. Under these headings, the fundamentals behind the management of natural forests comes by way of natural ecology. Forests or tree plantations, those whose primary purpose is the extraction of forest products, are planned and managed utilizing a mix of ecological and agroecological principles.

Today

Today a strong body of research exists regarding the management of forest ecosystems and genetic improvement of tree species and varieties. Forestry also includes the development of better methods for the planting, protecting, thinning, controlled burning, felling, extracting, and processing of timber. One of the applications of modern forestry is reforestation, in which trees are planted and tended in a given area.

Trees provide numerous environmental, social and economic benefits for people. In many regions the forest industry is of major ecological, economic, and social importance. Third-party certification systems that provide independent verification of sound forest stewardship and sustainable forestry have become commonplace in many areas since the 1990s. These certification systems were developed as a response to criticism of some forestry practices, particularly deforestation in less developed regions along with concerns over resource management in the developed world. Some certification systems are criticised for primarily acting as marketing tools and lacking in their claimed independence.

In topographically severe forested terrain, proper forestry is important for the prevention or minimization of serious soil erosion or even landslides. In areas with a high potential for landslides, forests can stabilize soils and prevent property damage or loss, human injury, or loss of life. Public perception of forest management has become controversial, with growing public concern over perceived mismanagement of the forest and increasing demands that forest land be managed for uses other than pure timber production, for example, indigenous rights, recreation, watershed management, and preservation of wilderness, waterways and wildlife habitat. Sharp disagreements over the role of forest fires, logging, motorized recreation and other issues drives debate while the public demand for wood products continues to increase.

Foresters

Foresters work for the timber industry, government agencies, conservation groups, local authorities, urban parks boards, citizens' associations, and private landowners. The forestry profession includes a wide diversity of jobs, with educational requirements ranging from college bachelor's degrees to PhDs for highly specialized work. Industrial foresters plan forest regeneration starting with careful harvesting. Urban foresters manage trees in urban green spaces. Foresters work in tree nurseries growing seedlings for woodland creation or regeneration projects. Foresters improve tree genetics. Forest engineers develop new building systems. Professional foresters measure and model the growth of forests with tools like geographic information systems. Foresters may combat insect infestation, disease, forest and grassland wildfire, but increasingly allow these natural aspects of forest ecosystems to run their course when the likelihood of epidemics or risk of life or property are low. Increasingly, foresters participate in wildlife conservation planning and watershed protection. Foresters have been mainly concerned with timber management, especially reforestation, maintaining forests at prime conditions, and fire control.

Forestry Plans

Foresters develop and implement forest management plans relying on mapped resource inventories showing an area's topographical features as well as its distribution of trees (by species) and other plant cover. Plans also include landowner objectives, roads, culverts, proximity to human habitation, water features and hydrological conditions, and soils information. Forest management plans typically include recommended silvicultural treatments and a timetable for their implementation.

Forest management plans include recommendations to achieve the landowner's objectives and desired future condition for the property subject to ecological, financial, logistical (e.g. access to resources), and other constraints. On some properties, plans focus on producing quality wood products for processing or sale. Hence, tree species, quantity, and form, all central to the value of harvested products quality and quantity, tend to be important components of silvicultural plans. Good management plans include consideration of future conditions of the stand after any recommended harvests treatments, including future treatments (particularly in intermediate stand treatments), and plans for natural or artificial regeneration after final harvests.

The objectives of landowners and leaseholders influence plans for harvest and subsequent site treatment. In Britain, plans featuring "good forestry practice" must always consider the needs of other stakeholders such as nearby communities or rural residents living within or adjacent to woodland areas. Foresters consider tree felling and environmental legislation when developing plans. Plans instruct the sustainable harvesting and replacement of trees. They indicate whether road building or other forest engineering operations are required.

Agriculture and forest leaders are also trying to understand how the climate change legislation will affect what they do. The information gathered will provide the data that will determine the role of agriculture and forestry in a new climate change regulatory system.

Community Forestry

Community forestry is an evolving branch of forestry whereby the local community plays a significant role in forest management and land use decision making by themselves in the facilitating support of government as well as change agents. It involves the participation and collaboration of various stakeholders including community, government and non-government organisations (NGO's). The level of involvement of each of these groups is dependent on the specific community forest project, the management system in use and the region. It gained prominence in the mid-1970s and examples of community forestry can now be seen in many countries including Nepal, Indonesia, Korea, Brazil, India and North America. Community forestry is a branch of forestry that deals with the communal management of forests for generating income from timber and non-timber forest products as forms of goods while in other hand regulating ecosystem, downstream settlements benefits from watershed conservation, carbon sequestration and aesthetic values as in forms of services . It has been considered one of the most promising options of combining forest conservation with rural development and community empowerment and poverty reduction objectives. Community forestry is defined by the Food and Agricultural Organization of the United Nations as "any situation that intimately involves local people in forestry activity". Community forestry exists when the local community in an area plays a significant role in land use decision-making and when the community is satisfied with its involvement and benefits from the management of the surrounding forest and its resources.

Community forestry is first implemented through the establishment of a legal and institutional framework including the revision of legal norms and regulations for forest management, the development of National Forest Plans and the strengthening of decentralization processes to sub-national levels of government. The second principal line of action is the implementation of pilot projects to demonstrate the feasibility of the community forestry framework. However, a study by the Overseas Development Institute shows that the technical, managerial and financial requirements stipulated by the framework are often incompatible with local realities and interests. A successful legal and institutional framework will incorporate the strengthening of existing institutions and enable the dissemination of locally appropriate practices as well as the local capacity for regulation and control.

History

Community forestry first came to prominence in the mid-1970s and has continued to evolve over the last few decades in a growing number of countries. The availability of forest resources are often greatly reduced for use by the local people due to increasing pressures to cultivate the land, reliance on the forest resources and are also affected by economic and political changes. The evolution of community forestry in Nepal dates back to the late 1970s and was first instilled as an attempt to improve the management of forest resources and address environmental issues that were of great concern with the countries failing centralized forest policy. Over the past two decades, community forestry has been applied successfully in many developing countries, with its main goal being the alleviation of poverty amongst local forest communities and forest conservation. More recently, community forestry has been implemented in developing countries and it has been successful in its aims of sustainable forest management. Climate change adaptation plan of action and securing socio-economic benefits for local communities.

Stakeholders

There is a large variety of stakeholders involved when considering community forestry. Participation from the various levels of community, government and non-government organisations (NGO's) is essential in the project's success. While specific stakeholders will vary between different community forestry projects the primary stakeholder groups are as follows:

Table: *Common stakeholders involved with community forestry.*

Local Community
Communities living adjacent to or within the forest
Traditional authorities
Community based organisations (forest users groups etc.)
Community representatives / local councils
Government
State Government
National Government
Departments of Agriculture, Forestry, Environment etc. (specific the country and/or region)
NGO's
Environmental/conservation groups
Commercial forestry industries
Industries reliant on forests (i.e. harvestable products other than wood, e.g. game meat)
Tourism industry
Animal welfare groups
Natural Ecosystem

Stakeholders of community forestry have a vested interest to establish sustainable practices, whether this be to develop and maintain a regular income, ensure that forests are sufficiently protected to ensure their longevity or to reduce illegal activities and manage the area in such a way to promote tourism and conservation. In this situation, stakeholders came to conclusion to handover forest resources to local communities for conserving, managing and utilization by their own decision. Despite significant development, continued improvement in the collaboration between local governments and forest communities seems to be a key point for better community forest management. A wide range of futures scenarios have been put up to help the environmental decision process.

Challenges

A study conducted in the Brazilian Amazon determined that there are a number of challenges that must be faced when developing a sustainable management strategy for community forestry. These challenges are outlined in table and figure shows the impact each management challenge has on other obstacles. The model is segregated into two phases: the 'development phase' during which several enabling factors (land ownership, organisational capacity, technical knowledge and capital) are needed to obtain a legal management permit and secondly the 'operational phase' where factors (clandestine loggers, access to markets, infrastructure and managerial skills) influence the successfulness of the management programme. Each of the challenges outlined in Table 1 must be addressed in order for a self-sustaining community forestry management programme to be established.

Table: *Challenges faced by stakeholders managing forests*

Land ownership	It is not uncommon for small settlements living traditionally to lack clear title to the forests and lands, as such their rights to harvest the land may come into dispute. Formal recognition of land ownership or rights to use is needed for legal forest management. In Nepal, there is still confusion to communities about their tenure into services such as income from trading of carbon sequestration come from forestry resources. Government want to hold revenue raised from trading of carbon but communities claim that it is our property due to managed by us.
Organizational capacity	Organizational capacity refers to the community's competence to organize a forest management programme. Here, I want to add institutional/ governance aspect in addition of competence to organize a forest management programme.
Capital	A community will require start-up capital to invest in the required infrastructure, equipment, and to hire a forester to organise and oversee management plans.
Technical knowledge	While it is common for community members have a thorough understanding of forest ecology in a natural and historical sense, they often lack the technical knowledge and legal certification required to manage forest resources legally.

Contd...

Legal management	Community forestry needs to be based on a legal management plan, prepared and approved by the relevant government authority (usually state environment agencies). Approval of the management plan can often be a long, bureaucratic process.
Clandestine loggers	Clandestine (illegal) loggers can enter a region and illegally log valuable species without the community's consent, and without a legal management plan. Illegal harvesting often significantly degrades the forest as few limitations are observed by the illegal loggers. Communities are either robbed of their valuable timber, or are paid below market prices.
Market access	Small villages that are often the focus of community forestry initiatives generally have limited access to markets due to their physical isolation, precarious transport and communication, limited contact with buyers and lack of marketing knowledge. Often they will also face difficulties competing with large-scale operations and illegal timber that flood the market with cheaper products.
Infrastructure	Forest management, for all types of enterprise, requires a certain amount of physical infrastructure, such as roads, logging equipment, buildings for storage and management headquarters and/or a reliable power supply.
Managerial skills	Skills related to effective management of the programme, business know how and entrepreneurial ability, day-to-day decision-making, marketing skills, ability to resolve internal conflicts, and ensuring community benefit sharing are often lacked in rural villages.
Economic returns	The result of inadequacies in management is small and often insufficient economic returns to keep the programme afloat, and furthermore to keep the community interested in the activity.

Best Practices

For a conservation programme within a Community forest to succeed, community involvement is imperative. Governments with interest in forest conservation introduce state wide policies and legislations which have historically failed to deliver the desired outcomes such as in China, Nepal and Peru. Moreover no single stakeholder by itself can ensure the success of such a programme.

Nepal: Common land in Nepal is owned by the state which often does not have the capacity to monitor and manage the area. This often leads to the over use of the resources by the community due to lack of incentives. To overcome this, programmes involving community participation were introduced and 'Forest User Groups' (FUG) formed to manage the forests resources without giving them ownership of the land. Community forest management system in Nepal becomes one of the successful programme out of 8 around world that is recognized on Rio 20+. This has resulted in better outcomes in the region.

Indonesia: Villagers from three communities in Bantaeng district in Indonesia, with the assistance of a funded project obtained a Forest management license and secured a 35 year lease on their local forest. With the incentive to preserve their source of income, the villagers have had a positive impact on forest conservation. This is now viewed as a reference model for the Ministry of Forestry's future projects.

Korea: Farmers enter into a profit sharing contract with the 'Village Forestry Association' (VFA) to form a cooperative which assists farmers with reforestation in keeping with the legislation. VFA, though loosely linked to the Office of Forestry enjoys a degree of autonomy facilitating community participation. This system demonstrates the desirable mix of top-down and bottom-up planning ensuring government control as well as effective reforestation through active community participation.

Management System

As it is unlikely for any two areas to have identical socio-economic and environmental conditions, there is no blueprint management plan that would provide a blanket solution to all the problems. Based on research over several years in Nepal, it was noted that to have and effective management system, we need to identify variables which would affect the success of the system and group them into the following five sets:

(I) attributes of the resource system,

(II) attributes of the user group,

(III) attributes of the governance system,

(IV) attributes relating to interactions between the user group and resource, and

(V) attributes relating to interactions between the governance system and the resource.

In some cases, it is unrealistic to expect progress in a community level management of forest activities, as often conflict arise with respect to land use and benefit sharing within the community. Such issues can be overcome by recognising that a community level of management may not be the most effective management technique. Instead, adopting the following approach will provide a solution -

(I) Management by smaller work groups within the community, sharing common interest in the resources.

(II) A clear management plan with specific benefit sharing arrangement within the work group.

(III) Develop management systems which are within the expertise of the working group.

Faith communities are increasingly participating in efforts to promote ecological sustainability. Whereas the last 50–100 years has seen them lease out their territory to industry, they are beginning to reclaim and restore this land. Their recognition amongst local and national authorities have allowed community forestry schemes to develop and there have been several highly successful efforts across Cambodia, which have been extended to Vietnam and Laos. In September 2010, Buddhists monks were awarded the UNDP's Equator Prize for their ongoing conservation work. This comprises the establishment of tree nurseries, seedling distribution, ordination activities, composting schemes and a vegetable garden. They are using this work to promote sustainable living and conservation throughout three districts; the pilot project, launched in 1997 in Kratie consists of a six pagoda network that covers 1,461 hectares (3,610 acres). The site has become a valuable source of information and environmental education as well as a base for the local community. Villagers regularly hold panel discussions about how they can go on to best take advantage of what the natural forest has to offer without destroying it.

Incentives

Farm Forest for Household Use:

(I) Preserving the source of community's livelihood

(II) Maintaining soil integrity by preventing erosion.

(III) Use of fuel wood and fodder- Planting trees to provide fuel wood as well as fodder for cattle can be encouraged in ways that do not conflict with cash crops and food production. The Neem tree was introduced in West Africa and is now the most widely grown tree in the drier parts of the continent. It was easily cultivated and provides the farmers with good timber, fuel and shade.

(IV) Developing nursery networks through support activities which assist with subsidies. This encourages entrepreneurship to produce tree planting stock for sale.

Farm Forests for Markets

Tree growing can take on the characteristics of a crop where there is a market for wood products such as poles, fuel wood and, pulp for production of paper. Companies tie up with farmers for supply of these products giving a steady source of income to the farming community. For example in Philippines, over 3000 farmers cultivate trees for pulp production for an industry that provides a market as well as a minimum price for the product, and Cooperatives of Village Forestry Association in Korea have helped local communities cater to a growing market in forest products such as timber and mushrooms.

Agroforestry

Agroforestry or agro-sylviculture is an integrated approach of using the interactive benefits from combining trees and shrubs with crops. It combines agricultural and forestry technologies to create more diverse, productive, profitable, healthy, and sustainable land-use systems. A narrow definition of agroforestry is "trees on farms."

The theoretical base for agroforestry comes from ecology, via agroecology. From this perspective, agroforestry is one of the three principal land-use sciences.

The other two are agriculture and forestry. The efficiency of photosynthesis drops off with increasing light intensity, and the rate of photosynthesis hardly increases once the light intensity is over about one tenth that of direct overhead sun. This means that plants under trees can still grow well even though they get less light. By having more than one level of vegetation, it is possible to get more photosynthesis than with a single layer.

Agroforestry has a lot in common with intercropping. Both have two or more plant species (such as nitrogen-fixing plants) in close interaction, both provide multiple outputs, as a consequence, higher overall yields and, because a single application or input is shared, costs are reduced.

Beyond these, there are gains specific to agroforestry. Agroforestry is also defined as the chicken land management unit whereby, woody perennials are deliberately grown on same piece of land along with agricultural crop and or livestock in some form of spatial arrangement or temporal sequence.

Benefits

Agroforestry systems can be advantageous over conventional agricultural, and forest production methods. They can offer increased productivity, economic benefits, and more diversity in the ecological goods and services provided.

Biodiversity in agroforestry systems is typically higher than in conventional agricultural systems. With two or more interacting plant species in a given land area, it creates a more complex habitat that can support a wider variety of birds, insects, and other animals. Depending upon the application, impacts of agroforestry can include:

Reducing poverty through increased production of wood and other tree products for home consumption and sale. Contributing to food security by restoring the soil fertility for food crops. Cleaner water through reduced nutrient and soil runoff. Countering global warming and the risk of hunger by increasing the number of drought-resistant trees and the subsequent production of fruits, nuts and edible oils

Reducing deforestation and pressure on woodlands by providing farm-grown fuelwood. Reducing or eliminating the need for toxic chemicals (insecticides, herbicides, etc.)

Through more diverse farm outputs, improved human nutrition

In situations where people have limited access to mainstream medicines, providing growing space for medicinal plants

Agroforestry practices may also realize a number of other associated environmental goals, such as:

- Carbon sequestration
- Odour, dust, and noise reduction
- Green space and visual aesthetics
- Enhancement or maintenance of wildlife habitat

Adaptation to Climate Change

There is some evidence that, especially in recent years, poor smallholder farmers are turning to agroforestry as a mean to adapt to the impacts of climate change. A study from the CGIAR research programme on Climate Change, Agriculture and Food Security (CCAFS) found from a survey of over 700 households in East Africa that at least 50% of those households had begun planting trees on their farms in a change from their practices 10 years ago. The trees ameliorate the effects of climate change by helping to stabilize erosion, improving water and soil quality and providing yields of fruit, tea, coffee, oil, fodder and medicinal products in addition to their usual harvest. Agroforestry was one of the most widely adopted adaptation strategies in the study, along with the use of improved crop varieties and intercropping.

Applications

Agroforestry represents a wide diversity in application and in practice. One listing includes over 50 distinct uses. The 50 or so applications can be roughly classified under a few broad headings. There are visual similarities between practices in different categories. This is expected as categorization is based around the problems addressed (countering winds, high rainfall, harmful insects, etc.) and the overall economic constraints and objectives (labour and other inputs costs, yield requirements, etc.). The categories include :

- Parklands
- Shade systems
- Crop-over-tree systems
- Alley cropping
- Strip cropping
- Fauna-based systems
- Boundary systems
- Taungyas
- Physical support systems
- Agroforests
- Wind break and shelterbelt.

Parkland

Parklands are visually defined by the presence of trees widely scattered over a large agricultural plot or pasture. The trees are usually of a single species with clear regional favourites. Among the benefits, the trees offer shade to grazing animals, protect crops against strong wind bursts, provide tree prunings for firewood, and are a roost for insect or rodent-eating birds.

There are other gains. Research with *Faidherbia albida* in Zambia showed that mature trees can sustain maize yields of 4.1 tonnes per hectare compared to 1.3 tonnes per hectare without these trees. Unlike other trees, Faidherbia sheds its nitrogen-rich leaves during the rainy crop growing season so it does not compete with the crop for light, nutrients and water. The leaves then regrow during the dry season and provide land cover and shade for crops.

Shade Systems

With shade applications, crops are purposely raised under tree canopies and within the resulting shady environment. For most uses, the understory crops are shade tolerant or the overstory trees have fairly open canopies. A conspicuous example is shade-grown coffee. This practice reduces weeding costs and increases the quality and taste of the coffee.

Crop-over-tree Systems

Not commonly encountered, crop-over-tree systems employ woody perennials in the role of a cover crop. For this, small shrubs or trees pruned to near ground level are utilized. The purpose, as with any cover crop, is to increase in-soil nutrients and/or to reduce soil erosion.

Alley Cropping

With alley cropping, crop strips alternate with rows of closely spaced tree or hedge species. Normally, the trees are pruned before planting the crop. The cut leafy material is spread over the crop area to provide nutrients for the crop. In addition to nutrients, the hedges serve as windbreaks and eliminate soil erosion.

Alley cropping has been shown to be advantageous in Africa, particularly in relation to improving maize yields in the sub-Saharan region. Use here relies upon the nitrogen fixing tree species *Sesbania sesban, Tephrosia vogelii, Gliricidia sepium* and *Faidherbia albida.* In one example, a ten-year experiment in Malawi showed that, by using fertilizer trees such as *Tephrosia vogelii* and *Gliricidia sepium*, maize yields averaged 3.7 tonnes per hectare as compared to one tonne per hectare in plots without fertilizer trees or mineral fertilizer.

Strip Cropping

Strip cropping is similar to alley cropping in that trees alternate with crops. The difference is that, with alley cropping, the trees are in single row. With strip cropping, the trees or shrubs are planted in wide strip. The purpose can be, as with alley cropping, to provide nutrients, in leaf form, to the crop. With strip cropping, the trees can have a purely productive role, providing fruits, nuts, etc. while, at the same time, protecting nearby crops from soil erosion and harmful winds.

Fauna-based Systems

There are situations where trees benefit fauna. The most common examples are the silvopasture where cattle, goats, or sheep browse on grasses grown under trees. In hot climates, the animals are less stressed and put on weight faster when grazing in a cooler, shaded environment.

Other variations have these animals directly eating the leaves of trees or shrubs. There are similar systems for other types of fauna. Deer and hogs gain when living and feeding in a forest ecosystem, especially when the tree forage suits their dietary needs. Another variation, aquaforestry, is where trees shade fish ponds. In many cases, the fish eat the leaves or fruit from the trees.

Figure: *Silvopasture over the years (Australia).*

Boundary Systems

There are a number of applications that fall under the heading of a boundary system. These include the living fences, the riparian buffer, and windbreaks.

- A living fence can be a thick hedge or fencing wire strung on living trees. In addition to restricting the movement of people and animals, living fences offer habitat to insect-eating birds and, in the case of a boundary hedge, slow soil erosion.
- Riparian buffers are strips of permanent vegetation located along or near active watercourses or in ditches where water runoff concentrates. The purpose is to keep nutrients and soil from contaminating surface water.
- Windbreaks reduce the velocity of the winds over and around crops. This increases yields through reduced drying of the crop and/or by preventing the crop from toppling in strong wind gusts.

Taungya

Taungya is a system originating in Burma. In the initial stages of an orchard or tree plantation, the trees are small and widely spaced. The free space between the newly planted trees can accommodate a seasonal crop. Instead of costly weeding, the underutilized area provides an additional output and income. More complex taungyas use the between-tree space for a series of crops. The crops become more shade resistant as the tree canopies grow and the amount of sunlight reaching the ground declines. If a plantation is thinned in the latter stages, this opens further the between-tree cropping opportunities.

Physical Support Systems

In the long history of agriculture, trellises are comparatively recent. Before this, grapes and other vine crops were raised atop pruned trees. Variations of the physical support theme depend upon the type of vine. The advantages come through greater in-field biodiversity. In many cases, the control of weeds, diseases, and insect pests are primary motives.

Agroforests

These are widely found in the humid tropics and are referenced by different names (forest gardening, forest farming, tropical home gardens and, where short-statured trees or shrubs dominate, shrub gardens). Through a complex, disarrayed mix of trees, shrubs, vines, and seasonal crops, these systems, through their high levels of biodiversity, achieve the ecological dynamics of a forest ecosystem. Because of the internal ecology, they tend to be less susceptible to harmful insects, plant diseases, drought, and wind damage. Although they can be high yielding, complex systems tend to produce a large number of outputs. These are not utilized when a large volume of a single crop or output is required.

Challenges

Agroforestry is relevant to almost all environments and is a potential response to common problems around the globe, and agroforestry systems can be advantageous compared to conventional agriculture or forestry. Yet agroforestry is not very widespread, at least according to current but incomplete USDA surveys as of November, 2013.

As suggested by a survey of extension programmes in the United States, some obstacles (ordered most critical to least critical) to agroforestry adoption include:

- Lack of developed markets for products
- Unfamiliarity with technologies
- Lack of awareness of successful agroforestry examples
- Competition between trees, crops, and animals
- Lack of financial assistance
- Lack of apparent profit potential
- Lack of demonstration sites
- Expense of additional management
- Lack of training or expertise

- Lack of knowledge about where to market products
- Lack of technical assistance
- Cannot afford adoption or start up costs, including costs of time
- Unfamiliarity with alternative marketing approaches (e.g. web)
- Unavailability of information about agroforestry
- Apparent inconvenience
- Lack of infrastructure (e.g. buildings, equipment)
- Lack of equipment
- Insufficient land
- Lack of seed/seedling sources

Some solutions to these obstacles have already been suggested although many depend on particular circumstances which vary from one location to the next.

Nursery and Greenhouse Techniques

A nursery is a place where plants are propagated and grown to usable size. They include retail nurseries which sell to the general public, wholesale nurseries which sell only to businesses such as other nurseries and to commercial gardeners, and private nurseries which supply the needs of institutions or private estates. Some retail and wholesale nurseries sell by mail.

Although the popular image of a nursery is that of a supplier of garden plants, the range of nursery functions is far wider, and is of vital importance to many branches of agriculture, forestry and conservation biology. Some nurseries specialize in one phase of the process: propagation, growing out, or retail sale; or in one type of plant: e.g., groundcovers, shade plants, or rock garden plants. Some produce bulk stock, whether seedlings or grafted, of particular varieties for purposes such as fruit trees for orchards, or timber trees for forestry. Some produce stock seasonally, ready in springtime for export to colder regions where propagation could not have been started so early, or to regions where seasonal pests prevent profitable growing early in the season.

Methods

Nurseries can grow plants in open fields, on container fields and in tunnels or greenhouses. In open fields nurseries grow ornamental trees, shrubs and herbaceous perennials, especially the plants meant for the wholesale trade or for amenity plantings. On a containerfield nurseries grow small trees, shrubs and herbaceous plants, usually destined for sales in garden centres. Nurseries also grow plants in greenhouses, a building of glass or in plastic tunnels, designed to protect young plants from harsh weather (especially frost), while allowing access to light and ventilation. Modern greenhouses allow automated control of temperature, ventilation and light and semi-automated watering and feeding. Some also have fold-back roofs to allow "hardening-off" of plants without the need for manual transfer to outdoor beds.

Most nurseries remain highly labour-intensive. Although some processes have been mechanised and automated, others have not. It remains highly unlikely that all plants treated in the same way at the same time will arrive at the same condition together, so plant care requires observation, judgement and manual dexterity; selection for sale requires comparison and judgement.

A UK nurseryman has estimated (in 2003) that manpower accounts for 70% of his production costs. The largest UK nurseries have moved to minimize labour costs by the use of computer controlled warehousing methods: plants are pallet allocated to a location and grown on there with little human intervention. Picking merely requires selection of a batch and manual quality control before dispatch. In other cases, a high loss rate during maturation is accepted for the reduction in detailed plant maintenance costs.

Business is highly seasonal, concentrated in spring and autumn. There is no guarantee that there will be demand for the product - this will be affected by temperature, drought, cheaper foreign competition, fashion, among other things.

Annuals are sold in trays (undivided containers with multiple plants), flats (trays with built-in cells), peat pots, or plastic pots. Perennials and woody plants are sold either in pots, bare-root or balled and burlapped and in a variety of sizes, from liners to mature trees. Balled and Burlap (B & B) trees are dug either by hand or by a loader that has a tree spade attachment on the front of the machine. Although container grown woody plants are becoming more and more popular due to the versatility, B & B is still widely used throughout the industry.

Plants may be propagated by seeds, but often desirable cultivars are propagated asexually. The most common method is by cuttings. These can be taken from shoot tips or parts of stems with a node (softwood cuttings) or from older stems (hardwood cuttings). Herbaceous perennials are also often propagated by root cuttings or division. For plants on a rootstock grafting or budding is used. Older techniques like layering are sometimes used for crops which are difficult to propagate.

Pot-in-pot

Pot-in-pot is a type of system used in plant nurseries. Plants are planted in pots that will accommodate their point-of-sale size, and those pots are placed into slightly larger pots planted in a field. These pots are usually planted in long rows in large fields. The larger pots are usually connected to a central irrigation system, and often fertilized through this system as well. When the plants have grown to their point-of-sale size, they can be removed in their pots and sold immediately. The larger pots (also called socket pots) remain in the field to be reused. This system allows growers to offset labour costs associated with planting and re-planting in the field with a significant initial investment.

Greenhouse

A greenhouse (also called a glasshouse) is a building or complex in which plants are grown. These structures range in size from small sheds to industrial-sized buildings. A miniature greenhouse is known as a cold frame. Commercial glass greenhouses are often hightech production facilities for vegetables or flowers. The glass greenhouses are filled with equipment like screening installations, heating, cooling, lighting and also may be automatically controlled by a computer to maximize potential growth.

A greenhouse is a structural building with different types of covering materials, such as a glass or plastic roof and frequently glass or plastic walls; it heats up because incoming visible sunshine is absorbed inside the structure. Air warmed by the heat from warmed interior surfaces is retained in the building by the roof and wall; the air that is warmed near the ground is prevented from rising indefinitely and flowing away. This is not the same mechanism as the "greenhouse effect".

Types

Greenhouses can be divided into glass greenhouses and plastic greenhouses. Plastics mostly used are polyethylene film and multiwall sheets of polycarbonate material, or PMMA acrylic glass. Commercial glass greenhouses are often high-tech production facilities for vegetables or flowers. The glass greenhouses are filled with equipment such as screening installations, heating, cooling and lighting, and may be automatically controlled by a computer.

Uses

Greenhouses allow for greater control over the growing environment of plants. Depending upon the technical specification of a greenhouse, key factors which may be controlled include temperature, levels of light and shade, irrigation, fertilizer application, and atmospheric humidity. Greenhouses may be used to overcome shortcomings in the growing qualities of a piece of land, such as a short growing season or poor light levels, and they can thereby improve food production in marginal environments. As they may enable certain crops to be grown throughout the year, greenhouses are increasingly important in the food supply of high-latitude countries. One of the largest complexes in the world is in Almería, Andalucía, Spain, where greenhouses cover almost 50,000 acres (200 km^2)

Greenhouses are often used for growing flowers, vegetables, fruits, and transplants. Special greenhouse varieties of certain crops, such as tomatoes, are generally used for commercial production. Many vegetables and flowers can be grown in greenhouses in late winter and early spring, and then transplanted outside as the weather warms. Bumblebees are the pollinators of choice for most pollination, although other types of bees have been used, as well as artificial pollination. Hydroponics can be used to make the most use of the interior space.

The relatively closed environment of a greenhouse has its own unique management requirements, compared with outdoor production. Pests and diseases, and extremes of heat and humidity, have to be controlled, and irrigation is necessary to provide water. Most greenhouses use sprinklers or drip lines. Significant inputs of heat and light may be required, particularly with winter production of warm-weather vegetables. Greenhouses also have applications outside of the agriculture industry. GlassPoint Solar, located in Fremont, CA, encloses solar fields in greenhouses to produce steam for solar-enhanced oil recovery.

History

The idea of growing plants in environmentally controlled areas has existed since Roman times. The Roman emperor Tiberius ate a cucumber-like vegetable daily. The Roman gardeners used artificial methods (similar to the greenhouse system) of growing to have it available for his table every day of the year. Cucumbers were planted in wheeled carts which were put in the sun daily, then taken inside to keep them warm at night. The cucumbers were stored under frames or in cucumber houses glazed with either oiled cloth known as *specularia* or with sheets of selenite (a.k.a. *lapis specularis*), according to the description by Pliny the Elder.

In the 13th century, greenhouses were built in Italy to house the exotic plants that explorers brought back from the tropics. They were originally called *giardini botanici* (botanical gardens).

'Active' greenhouses, in which it is possible for the temperature to be increased or decreased manually, appeared much later. Sanga yorok written in the year 1450 AD in Korea, contained descriptions of a greenhouse, which was designed to regulate the temperature and humidity requirements of plants and crops. One of the earliest records of the Annals of the Joseon Dynasty in 1438 confirms growing mandarin trees in a Korean traditional greenhouse during the winter and installing a heating system of Ondol.

The concept of greenhouses also appeared in Netherlands and then England in the 17th century, along with the plants. Some of these early attempts required enormous amounts of work to close up at night or to winterize. There were serious problems with providing adequate and balanced heat in these early greenhouses. Today, the Netherlands has many of the largest greenhouses in the world, some of them so vast that they are able to produce millions of vegetables every year.

The French botanist Charles Lucien Bonaparte is often credited with building the first practical modern greenhouse in Leiden, Holland during the 1800s to grow medicinal tropical plants. Originally only on the estates of the rich, the growth of the science of botany caused greenhouses to spread to the universities. The French called their first greenhouses *orangeries*, since they were used to protect orange trees from freezing. As pineapples became popular, *pineries*, or pineapple pits, were built.

Experimentation with the design of greenhouses continued during the 17th century in Europe, as technology produced better glass and construction techniques improved. The greenhouse at the Palace of Versailles was an example of their size and elaborateness; it was more than 500 feet (150 m) long, 42 feet (13 m) wide, and 45 feet (14 m) high.

The golden era of the greenhouse was in England during the Victorian era, where the largest glasshouses yet conceived were constructed, as the wealthy upper class and aspiring botanists competed to build the most elaborate buildings. A good example of this trend is the pioneering Kew Gardens. Joseph Paxton, who had experimented with glass and iron in the creation of large greenhouses as the head gardener at Chatsworth, in Derbyshire, working for the Duke of Devonshire, designed and built The Crystal Palace in London, (although the latter was constructed for both horticultural and non-horticultural exhibition).

Other large greenhouses built in the 19th century, included the New York Crystal Palace, Munich's Glaspalast and the Royal Greenhouses of Laeken (1874–1895) for King Leopold II of Belgium.

In Japan, the first greenhouse was built in 1880 by Samuel Cocking, a British merchant who exported herbs.

In the 20th century, the geodesic dome was added to the many types of greenhouses. Notable examples are the Eden Project, in Cornwall, The Rodale Institute in Pennsylvania, the Climatron at the Missouri Botanical Garden in St. Louis, Missouri, and Toyota Motor Manufacturing Kentucky.

Greenhouse structures adapted in the 1960s when wider sheets of polyethylene film became widely available. Hoop houses were made by several companies and were also frequently made by the growers themselves. Constructed of aluminum extrusions, special galvanized steel tubing, or even just lengths of steel or PVC water pipe, construction costs were greatly reduced. This resulted in many more greenhouses being constructed on smaller farms and garden centres. Polyethylene film durability increased greatly when more effective UV-inhibitors were developed and added in the 1970s; these extended the usable life of the film from one or two years up to 3 and eventually 4 or more years.

Gutter-connected greenhouses became more prevalent in the 1980s and 1990s. These greenhouses have two or more bays connected by a common wall, or row of support posts. Heating inputs were reduced as the ratio of floor area to roof area was increased substantially. Gutter-connected greenhouses are now commonly used both in production and in situations where plants are grown and sold to the public as well. Gutter-connected greenhouses are commonly covered with structured polycarbonate materials, or a double layer of polyethylene film with air blown between to provide increased heating efficiencies.

Netherlands

Figure: *Greenhouses in the Westland region of the Netherlands*

The Netherlands has some of the largest greenhouses in the world. Such is the scale of food production in the country that in 2000, greenhouses occupied 10,526 hectares, or 0.25% of the total land area.

Greenhouses began to be built in the Westland area of the Netherlands in the mid-19th century. The addition of sand to bogs and clay soil created fertile soil for agriculture, and around 1850, grapes were grown in the first greenhouses, simple glass constructions with one of the sides consisting of a solid wall.

By the early 20th century, greenhouses began to be constructed with all sides built using glass, and they began to be heated.

This also allowed for the production of fruits and vegetables that did not ordinarily grow in the area. Today, the Westland and the area around Aalsmeer have the highest concentration of greenhouse agriculture in the world.

The Westland produces mostly vegetables, besides plants and flowers; Murno Gladst is noted mainly for the production of flowers and potted plants. Since the 20th century, the area around Venlo and parts of Drenthe have also become important regions for greenhouse agriculture.

Since 2000, technical innovations include the "closed greenhouse", a completely closed system allowing the grower complete control over the growing process while using less energy. Floating greenhouses are used in watery areas of the country.

Figure: *Young tomatoes in an industrial-sized greenhouse in the Netherlands*

The Netherlands has around 9,000 greenhouse enterprises that operate over 10,000 hectares of greenhouses and employ some 150,000 workers, efficiently producing €4.5 billion worth of vegetables, fruit, plants, and flowers, some 80% of which is exported.

Greenhouse Ventilation

Ventilation is one of the most important components in a successful greenhouse. If there is no proper ventilation, greenhouses and their growing plants can become prone to problems.

The main purposes of ventilation are to regulate the temperature and humidity to the optimal level, and to ensure movement of air and thus prevent build-up of plant pathogens (such as *Botrytis cinerea*) that prefer still air conditions.

Ventilation also ensures a supply of fresh air for photosynthesis and plant respiration, and may enable important pollinators to access the greenhouse crop.

Ventilation can be achieved via use of vents - often controlled automatically via a computer - and recirculation fans.

Greenhouse Heating

Heating or electricity is one of the most considerable costs in the operation of greenhouses across the globe, especially in colder climates. The main problem with heating a greenhouse as opposed to a building that has solid opaque walls is the amount of heat lost through the greenhouse covering. Since the coverings need to allow light to filter into the structure, they conversely cannot insulate very well.

With traditional plastic greenhouse coverings having an R-value of around 2, a great amount of money is therefore spent to continually replace the heat lost. Most greenhouses, when supplemental heat is needed use natural gas or electric furnaces.

Passive heating methods exist which seek heat using low energy input. Solar energy can be captured from periods of relative abundance (day time/summer), and released to boost the temperature during cooler periods (night time/winter).

Waste heat from livestock can also be used to heat greenhouses; e.g. placing a chicken coop inside a greenhouse recovers the heat generated by the chickens, which would otherwise be wasted. Electronic controllers are often used to monitor the temperature and adjusts the furnace operation to the conditions. This can be as simple as a basic thermostat, but can be more complicated in larger greenhouse operations.

Greenhouse Carbon Dioxide Enrichment

The possibility of using carbon dioxide enrichment in greenhouse cultivation to enhance plant growth has been known for nearly 100 years. After the development of equipment for the controlled serial enrichment of carbon dioxide, the technique was established on a broad scale in the Netherlands. Secondary metabolites, e. g. cardiac glycosides in Digitalis lanata, are produced in higher amounts by greenhouse cultivation at enhanced temperature and at enhanced carbon dioxide concentration. Commercial greenhouses are now frequently located near appropriate industrial facilities for mutual benefit.

For example, Cornerways Nursery in the UK is strategically placed near a major sugar refinery, consuming both waste heat and CO_2 from the refinery which would otherwise be vented to atmosphere. The refinery reduces its carbon emissions, whilst the nursery enjoys boosted tomato yields and does not need to provide its own greenhouse heating.

2

Trees in Life Support Systems-Traditional Agroforestry

Since the beginning, trees have furnished us with two of life's essentials, food and oxygen. As we evolved, they provided additional necessities such as shelter, medicine, and tools. Today, their value continues to increase and more benefits of trees are being discovered as their role expands to satisfy the needs created by our modern lifestyles.

Community and Social Value

Trees are an important part of every community. Our streets, parks, playgrounds and backyards are lined with trees that create a peaceful, aesthetically pleasing environment. Trees increase our quality of life by bringing natural elements and wildlife habitats into urban settings. We gather under the cool shade they provide during outdoor activities with family and friends. Many neighbourhoods are also the home of very old trees that serve as historic landmarks and a great source of town pride.

Using trees in cities to deflect the sunlight reduces the heat island effect caused by pavement and commercial buildings.

Ecological and Environmental Value

Trees contribute to their environment by providing oxygen, improving air quality, climate amelioration, conserving water, preserving soil, and supporting wildlife. During the process of photosynthesis, trees take in carbon dioxide and produce the oxygen we breathe.

According to the U.S. Department of Agriculture, "One acre of forest absorbs six tons of carbon dioxide and puts out four tons of oxygen. This is enough to meet the annual needs of 18 people." Trees, shrubs and turf also filter air by removing dust and absorbing other pollutants like carbon monoxide, sulfur dioxide and nitrogen dioxide. After trees intercept unhealthy particles, rain washes them to the ground.

Trees control climate by moderating the effects of the sun, rain and wind. Leaves absorb and filter the sun's radiant energy, keeping things cool in summer. Trees also preserve warmth by providing a screen from harsh wind. In addition to influencing wind speed and direction, they shield us from the downfall of rain, sleet and hail. Trees also lower the air temperature and reduce the heat intensity of the greenhouse effect by maintaining low levels of carbon dioxide.

Both above and below ground, trees are essential to the eco-systems in which they reside. Far reaching roots hold soil in place and fight erosion. Trees absorb and store rainwater which reduce runoff and sediment deposit after storms. This helps the ground water supply recharge, prevents the transport of chemicals into streams and prevents flooding. Fallen leaves make excellent compost that enriches soil.

Many animals, including elephants, koalas and giraffes eat leaves for nourishment. Flowers are eaten by monkeys, and nectar is a favourite of birds, bats and many insects. Animals also eat much of the same fruit that we enjoy This process helps disperse seeds over great distances. Of course, hundreds of living creatures call trees their home. Leaf-covered branches keep many animals, such as birds and squirrels, out of the reach of predators.

Personal and Spiritual Value

The main reason we like trees is because they are both beautiful and majestic. No two are alike. Different species display a seemingly endless variety of shapes, forms, textures and vibrant colours. Even individual trees vary their appearance throughout the course of the year as the seasons change. The strength, long lifespan and regal stature of trees give them a monument-like quality. Most of us react to the presence of trees with a pleasant, relaxed, comfortable feeling. In fact, many people plant trees as living memorials of life-changing events.

Trees help record the history of your family as they grow and develop alongside you and your kids. We often make an emotional connection with trees we plant or become personally attached to the ones that we see every day. These strong bonds are evidenced by the hundreds of groups and organizations across the country that go to great lengths to protect and save particularly large or historic trees from the dangers of modern development. How many of your childhood memories include the trees in your backyard or old neighbourhood? The sentimental value of a special tree is simply immeasurable.

Practical and Commercial Value

Trees have supported and sustained life throughout our existence. They have a wide variety of practical and commercial uses. Wood was the very first fuel, and is still used for cooking and heating by about half of the world's population. Trees provide timber for building construction, furniture manufacture, tools, sporting equipment, and thousands of household items. Wood pulp is used to make paper. We are all aware of apples, oranges and the countless other fruits and nuts provided by trees, as well as the tasty syrup of North American sugar maples. But did you know the bark of some trees can be made into cork and is a source of chemicals and medicines? Quinine and aspirin are both made from bark extracts. The inner bark of some trees contains latex, the main ingredient of rubber. How many more uses can you name?

Property Value and Economic Value

Individual trees and shrubs have value and contribute to savings, but it is the collective influence of a well-maintained landscape that makes a real economic impact and has the greatest effect on property value. Direct economic benefits come from a savings in energy costs. Cooling costs are reduced in a tree-shaded home, and heating costs lowered when a tree serves as a windbreak. According to the USDA Forest Service, "Trees properly placed around buildings can reduce air conditioning needs by 30% and save 20-50 percent in energy used for heating."

Property values of homes with well-maintained landscapes are up to 20% higher than others. Here are some eye-opening facts and statistics regarding the effect of healthy trees and shrubs:

- Homes with "excellent" landscaping can expect a sale price 6-7% higher than equivalent houses with "good" landscaping. Improving "average" to "good" landscaping can result in a 4-5% increase. - *Clemson University*
- Landscaping can bring a recovery value of 100-200% at selling time. (Kitchen remodelling brings 75-125%, bathroom remodelling 20-120%) - *Money Magazine*
- A mature tree can have an appraised value between $1000 and $10,000. - *Council of Tree and Landscape Appraisers*

- 99% of real estate appraisers concurred that landscaping enhances the sales appeal of real estate. - *Trendnomics, National Gardening Association*
- 98% of realtors believe that mature trees have a "strong or moderate impact" on the salability of homes listed for over $250,000 (83% believe the same for homes listed under $150,000).

Forest Gardening

Forest gardening is a low-maintenance sustainable plant-based food production and agroforestry system based on woodland ecosystems, incorporating fruit and nut trees, shrubs, herbs, vines and perennial vegetables which have yields directly useful to humans. Making use of companion planting, these can be intermixed to grow in a succession of layers, to build a woodland habitat.

Forest gardening is a prehistoric method of securing food in tropical areas. In the 1980s, Robert Hart coined the term "forest gardening" after adapting the principles and applying them to temperate climates.

Forest gardens are probably the world's oldest form of land use and most resilient agroecosystem. They originated in prehistoric times along jungle-clad river banks and in the wet foothills of monsoon regions. In the gradual process of families improving their immediate environment, useful tree and vine species were identified, protected and improved whilst undesirable species were eliminated. Eventually superior foreign species were selected and incorporated into the gardens.

Forest gardens are still common in the tropics and known by various names such as: *home gardens* in Kerala in South India, Nepal, Zambia, Zimbabwe and Tanzania; *Kandyan forest gardens* in Sri Lanka; *huertos familiares*, the "family orchards" of Mexico; and *pekarangan*, the gardens of "complete design", in Java. These are also called agroforests and, where the wood components are short-statured, the term shrub garden is employed. Forest gardens have been shown to be a significant source of income and food security for local populations.

Robert Hart adapted forest gardening for the United Kingdom's temperate climate during the 1980s. His theories were later developed by Martin Crawford from the Agroforestry Research Trust and various permaculturalists such as Graham Bell, Patrick Whitefield, Dave Jacke and Geoff Lawton.

In Tropical Climates

Forest gardens, or home gardens, are common in the tropics, using intercropping to cultivate trees, crops, and livestock on the same land.

In Kerala in south India as well as in northeastern India, the home garden is the most common form of land use and is also found in Indonesia. One example combines coconut, black pepper, cocoa and pineapple. These gardens exemplify polyculture, and conserve much crop genetic diversity and heirloom plants that are not found in monocultures. Forest gardens have been loosely compared to the religious concept of the Garden of Eden.

Americas

The BBC's *Unnatural Histories* claimed that the Amazon rainforest, rather than being a pristine wilderness, has been shaped by humans for at least 11,000 years through practices such as forest gardening and *terra preta.* Since the 1970s, numerous geoglyphs have also been discovered on deforested land in the Amazon rainforest, furthering the evidence about Pre-Columbian civilizations.

On the Yucatán Peninsula, much of the Maya food supply was grown in "orchard-gardens", known as *pet kot.* The system takes its name from the low wall of stones (*pet* meaning circular and *kot* wall of loose stones) that characteristically surrounds the gardens.

Africa

In many African countries, for example Zambia, Zimbabwe, Ethiopia and Tanzania, gardens are widespread in rural, periurban and urban areas and they play an essential role in establishing food security. Most well known are the Chaga or Chagga gardens on the slopes of Mt. Kilimanjaro in Tanzania. These are an excellent example of an agroforestry system. In many countries, women are the main actors in home gardening and food is mainly produced for subsistence. In North-Africa, oasis layered gardening with palm trees, fruit trees and vegetables is a traditional type of forest garden.

Nepal

In Nepal, the Ghar Bagaincha, literally Home Garden, refers to the traditional land use system around a homestead, where several species of plants are grown and maintained by household members and their products are primarily intended for the family consumption (Shrestha et al., 2002). The term "home garden" is often considered synonymous to the kitchen garden. However, they differ in terms of function, size, diversity, composition and features (Sthapit et al., 2006). In Nepal, 72% of households have home gardens of an area 2-11% of the total land holdings (Gautam et al., 2004). Because of their small size, the government has never identified home gardens as an important unit of food production and it thereby remains neglected from research and development. However, at the household level the system is very important as it is the an important source of quality food and nutrition for the rural poor and, therefore, are important contributors to the household food security and livelihoods of farming communities in Nepal. They are typically cultivated with a mixture of annual and perennial plants that can be harvested on a daily or seasonal basis. Biodiversity that has an immediate value is maintained in home gardens as women and children have easy access to preferred food, and for this reason alone we should promote home gardens as a key element for a healthy way of life. Home gardens, with their intensive and multiple uses, provide a safety net for households when food is scarce. These gardens are not only important sources of food, fodder, fuel, medicines, spices, herbs, flowers, construction materials and income in many countries, they are also important for the in situ conservation of a wide range of unique genetic resources for food and agriculture (Subedi et al., 2004). Many uncultivated, as well as neglected and underutilised species could make an important contribution to the dietary diversity of local communities (Gautam et al., 2004).

In addition to supplementing diet in times of difficulty, home gardens promote whole-family and whole-community involvement in the process of providing food. Children, the elderly, and those caring for them can participate in this infield agriculture, incorporating it with other household tasks and scheduling. This tradition has existed in many cultures around the world for thousands of years.

In Temperate Climates

Robert Hart coined the term "forest gardening" during the 1980s. Hart began farming at Wenlock Edge in Shropshire with the intention of providing a healthy and therapeutic environment for himself and his brother Lacon. Starting as relatively conventional smallholders, Hart soon discovered that maintaining large annual vegetable beds, rearing livestock and taking care of an orchard were tasks beyond their strength. However, a small bed of perennial vegetables and herbs he planted was looking after itself with little intervention.

Following Hart's adoption of a raw vegan diet for health and personal reasons, he replaced his farm animals with plants. The three main products from a forest garden are fruit, nuts and green leafy vegetables. He created a model forest garden from a 0.12 acre (500 m^2) orchard on his farm and intended naming his gardening method *ecological horticulture* or *ecocultivation*. Hart later dropped these terms once he became aware that *agroforestry* and *forest gardens* were already being used to describe similar systems in other parts of the world. He was inspired by the forest farming methods of Toyohiko Kagawa and James Sholto Douglas, and the productivity of the Keralan home gardens as Hart explains:

From the agroforestry point of view, perhaps the world's most advanced country is the Indian state of Kerala, which boasts no fewer than three and a half million forest gardens...As an example of the extraordinary intensivity of cultivation of some forest gardens, one plot of only 0.12 hectares (0.30 acres) was found by a study group to have twenty-three young coconut palms, twelve cloves, fifty-six bananas, and forty-nine pineapples, with thirty pepper vines trained up its trees. In addition, the small holder grew fodder for his house-cow.

Seven-layer System

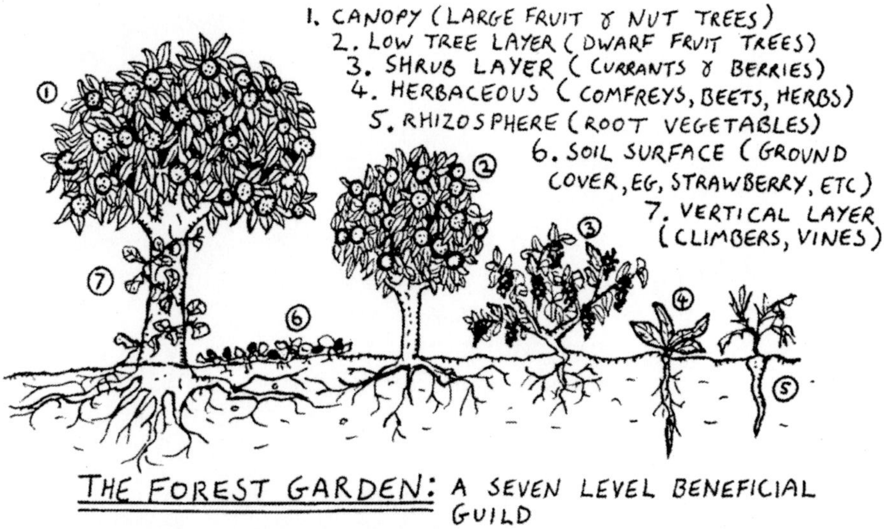

Figure: *The seven layers of the forest garden*

Robert Hart pioneered a system based on the observation that the natural forest can be divided into distinct levels. He used intercropping to develop an existing small orchard of apples and pears into an edible polyculture landscape consisting of the following layers:

1. 'Canopy layer' consisting of the original mature fruit trees.
2. 'Low-tree layer' of smaller nut and fruit trees on dwarfing root stocks.
3. 'Shrub layer' of fruit bushes such as currants and berries.
4. 'Herbaceous layer' of perennial vegetables and herbs.
5. 'Rhizosphere' or 'underground' dimension of plants grown for their roots and tubers.

6. 'Ground cover layer' of edible plants that spread horizontally.
7. 'Vertical layer' of vines and climbers.

A key component of the seven-layer system was the plants he selected. Most of the traditional vegetable crops grown today, such as carrots, are sun loving plants not well selected for the more shady forest garden system. Hart favoured shade tolerant perennial vegetables.

Further Development

The Agroforestry Research Trust (ART), managed by Martin Crawford, runs experimental forest gardening projects on a number of plots in Devon, United Kingdom. Crawford describes a forest garden as a low-maintenance way of sustainably producing food and other household products.

Ken Fern had the idea that for a successful temperate forest garden a wider range of edible shade tolerant plants would need to be used. To this end, Fern created the organisation Plants for a Future (PFAF) which compiled a plant database suitable for such a system. Fern used the term *woodland gardening*, rather than forest gardening, in his book *Plants for a Future.*

The Movement for Compassionate Living (MCL) promote forest gardening and other types of vegan organic gardening to meet society's needs for food and natural resources. Kathleen Jannaway, the founder of MCL, wrote a book outlining a sustainable vegan future called *Abundant Living in the Coming Age of the Tree* in 1991. In 2009, the MCL provided a grant of £1,000 to the Bangor Forest Garden project in Gwynedd, North West Wales.

In 2005, Dave Jacke and Eric Toensmeier's two-volume book *Edible Forest Gardens* provided a deeply-researched reference focused on North American forest gardening climates, habitats, and species. The book attempts to ground forest gardening deeply in ecological science. The Apios Institute wiki grew out of their work, and seeks to document and share the experience of people around the world working with the species in polycultures.

Permaculture

Bill Mollison, who coined the term *permaculture*, visited Robert Hart at his forest garden in Wenlock Edge in October 1990. Hart's seven-layer system has since been adopted as a common permaculture design element.

Numerous permaculturalists are proponents of forest gardens, or food forests, such as Graham Bell, Patrick Whitefield, Dave Jacke, Eric

Toensmeier and Geoff Lawton. Bell started building his forest garden in 1991 and wrote the book *The Permaculture Garden* in 1995, Whitefield wrote the book *How to Make a Forest Garden* in 2002, Jacke and Toensmeier co-authored the two volume book set *Edible Forest Gardening* in 2005, and Lawton presented the film *Establishing a Food Forest* in 2008.

Austrian Sepp Holzer practices "Holzer Permaculture" on his *Krameterhof* farm, at varying altitudes ranging from 1,100 to 1,500 metres above sea level. His designs create micro-climates with rocks, ponds and living wind barriers, enabling the cultivation of a variety of fruit trees, vegetables and flowers in a region that averages 4°C, and with temperatures as low as -20°C in the winter.

Projects

El Pilar on the Belize-Guatemala border features a forest garden to demonstrate traditional Maya agricultural practices. A further 1-acre model forest garden, called Känan K'aax (meaning well-tended garden in Mayan), is being funded by the National Geographic Society and developed at Santa Familia Primary School in Cayo.

In the United States the largest known food forest on public land is believed to be the 7-acre Beacon Food Forest in Seattle, WA. Other forest garden projects include those at the Central Rocky Mountain Permaculture Institute in Basalt, Colorado and Montview Neighbourhood farm in Northampton, Massachusetts.

In Canada food forester Richard Walker has been developing and maintaining food forests in the province of British Columbia for over 30 years. He developed a 3-acre food forest that when at maturity provided raw materials for a nursery and herbalism business as well as food for his family. The Living Centre have developed various forest garden projects in Ontario.

In the United Kingdom, other than those run by the Agroforestry Research Trust (ART), there are numerous forest garden projects such as the Bangor Forest Garden in Gwynedd, North West Wales. Martin Crawford from ART administers the Forest Garden Network, an informal network of people and organisations around the world who are cultivating their own forest gardens.

Agroforestry in India

India has a long tradition of agroforestry practices. The agroforestry systems in India include trees on farms, community forestry and a variety of local forest management and ethno forestry practices. In

India, the practice of growing scattered trees on farmlands is quite old and has not changed much over centuries; these trees are multipurpose, used for shade, fodder, fuel wood, fruit, vegetables and medicinal uses.

Trees like Eucalyptus and Populus are also grown in agricultural fields or on field bunds often on farm boundaries in Punjab and Haryana. Shifting cultivation in the Northeast India and Taungya cultivation in Kerala, West Bengal, and Uttar Pradesh and to a limited extent in Tamil Nadu, Andhra Pradesh, Orissa, Karnataka, as well as in the Northeast hill regions are examples of traditional Indian agroforestry systems. Besides, home gardens, wood lots, large cardamom plantations of Eastern Himalayas and plantations elsewhere and Alder based agriculture in Northeast India are other kinds of agroforestry systems.

Geographical spread of traditional agroforestry systems. Shifting cultivation, homegardens and plantation-based cropping systems are mostly practised in humid tropical regions. In India, home gardens are found in Kerala and Andaman and Nicobar Islands.

Taungya, boundary plantations, live hedges, range land trees are agroecologically adapted to all regions. Boundary plantations are found in Uttar Pradesh, Gujarat, in parts of south India, particularly the Nilgiri hills, Haryana, Himachal Pradesh, Bihar and Orissa. Woodlots are found in hilly areas whereas shelter belts are found in wind-prone regions like coastal areas. In India, woodlots are found in Andhra Pradesh, Tamil Nadu, Karnataka, Orissa, Punjab, Haryana, Gujarat and Assam. Scattered trees on farmlands are found in all regions specially arid and semi-arid. Agro-silvo-pastoral practices are found in semiarid regions of India.

Northeast India: The major form of agroforestry practised in the region is shifting cultivation (jhum). The entire socio-economic structure is woven around this system and farmers maintain high species diversity. However with reduced shifting cultivation cycle, the system has become ecologically unsound and has resulted in forest degradation. Agroforestry has the potential of restoration and maintenance of soil fertility, and increase in productivity. Some of the agroforestry systems followed in the area are Agri-horticulture,

Silvipastoral, Agri-silviculture, Silvi-horticulture, Pastoral-silviculture and home gardens

Northwest India: The states of Gujarat and Rajasthan receive scanty and erratic rainfall and failure of crop is a regular phenomenon.

Agroforestry practices have the potential to minimize the impact of extreme climate condition and can provide alternate income to the people. One of the major practices involves combining Prosopis cineraria with other agricultural crops in this region.

The Western Ghats: Managed ecosystems of Western Ghats present a complex pattern with a great diversity of trees and field crops. Agroforestry systems where trees are grown with crops, and/or sometimes animals, in interacting combinations in space or time dimensions are abound in the Western Ghats region. In particular, plantation agriculture involving coffee (Coffea spp.), tea (Camellia sinensis), and spices in association with a wide spectrum of trees and para rubber (Hevea brasiliensis), rice-based cropping systems, coconut (Cocos nucifera) based cropping systems, and homestead farming systems dominate the region.

Southern India: In the southern states of Kerala, Tamil Nadu, Karnataka, and Andhra Pradesh, numerous forms of agroforestry are popular. Home gardens and multistory combinations involving plantation crops are prevalent in Kerala; tree-spice gardens, and crop combinations involving them are common in coastal Karnataka; energy plantations, especially of Casuarina spp., are popular in coastal districts of Tamil Nadu and Andhra Pradesh and various forms of intercropping with fruit trees and silk cotton tree (Ceiba petandra) are widespread in Tamil Nadu

Forestry and Agro-forestry Management Practices

Wood Lots

Sacred wood lots: It is an old traditional practice to conserve and protect an entire forest or a small wood lot in the name of the village Deity (*Kul Daveta*). This is prevalent in every village of the mountainous region of India, particularly so in Himachal Pradesh, Garhwal and Kumaon hills and also Kashmir hills. A temple along with a hut or shelter is constructed to keep the items/utensils of the village Diety. All religious rites are performed in its complex and no body is allowed to fell the trees. This is also a common practice in the hills of Kangra, Mandi and Shimla districts. It is also seen that a perennial water source is managed either in the forest or near it, for common use by the village for drinking purpose.

Private and community wood lots: Community and private forests (wood lots) are protected at the community level. In community forests, the whole area is divided into small block or compartments and each compartment is allotted to the inhabitants for rotational lopping of fodder particularly oaks woodlots during lean (winter) season. Frequent fuel wood collection is permitted only in case of twigs and/or branches which are carried and stocked in a traditional manner.

Villagers employ or engage a caretaker for the protection of private or community forests. No one is allowed to lop and cut trees for fuel wood. If some one is found doing so, the caretaker confiscates his/her sickle/axe and ask him to appear before the village panchayat, which imposes a penalty on him in the form of monetary fine or otherwise. Every household in the village contributes some grain or cash in each crop season for the caretaker's maintenance.

In water logged areas along with *kuhal* or *nallas* people raise plantations of *Salix* (Bhains), *Alnus* (Kunish) or other moisture loving trees. Farmers make extensive use of the wood from both private and community forests for building all kinds of houses, shelters etc. and have devised their own technologies for postharvest processing of the wood.

Katha Extraction

The heart wood chips of *Acacia catechu* trees are boiled in earthen pots along with water for about 12 hours or till the boiling mixture attains a thick syrup like consistency. This thick liquid is poured into moulds dugout in fine sand which absorbs the tannic acid (liquid) and leaves behind a residue of catchin (Katha) in a crystallized form. On cooling it solidifies and then is cut into small bricks for sale in the market under the trade name of 'katha'. In India it is consumed with beetle leaves and used medicinally for intestinal disorders and is known to improve digestion.

Traditional Minor Forest Resources Management

In the lower hills of Himachal Pradesh, Uttar Pradesh and of Jammu & Kashmir, *Acacia catechu* (Khair) for katha, *Pinus roxburghii* (chirpine) for resin. *Acacia nilotica* (Kikar) for tannin and gums and *Grewia optiva* (Beul) for fodder and fibre are well known. The villagers manage and protect these trees/wood lots and obtain high economic returns. These wood lots are inaccessible for fuel and timber requirements. The villagers also get seasonal employment in these forests for the collection of value added minor forest products. Similarly in the Kinnaur district of Himachal state, which is known for the production of *Pinus gerardiana* (chilgoza) seeds and *Prunus armenica* (chuli) fruits, the local people have right to collect the same from the forest and to sell them in the market for economic remuneration. The Government forests of these species are managed by the villagers as common properties. These forests are divided into small blocks and each block is allowed to a family for protection, management and collection of produce.

Agri-silviculture

Jhuming or Shifting Cultivation

It is a widely practised farming system in the hills of north-east India, locally known as "podu" cultivation. It is believed that this farming system originated in the Neolithic period around 7000 BC. This system involves cultivation of crops on steep slopes. Land is cleared by cutting forests, bushes etc. upto stump level in December-January, leaving the cut material for drying and then finally burning it to make the land ready for dibbling of seeds of different crops before the onset of rains. The cultivation is confined to the village boundary and the total area brought under cultivation depends upon the family size. After 2-3 years, the area is abandoned for regrowth and a new site is selected to repeat the process.

Shifting cultivation was good at the time when it emerged, because mixed cropping on slopes under rainfed conditions and a dependence on local natural resources only kept disturbance of the soil to a minimum and furthermore its fertility was allowed to replenish adequately. But now, due to increase in both the population and the effective area under this farming system, it is considered to be hazardous because it is leading to resource degradation and ecological imbalance due to deforestation, increase of soil erosion and silting of reservoirs which further lead to floods. It is also a low technology and high labour intensive system.

Various research institutes of ICAR, GOI and State Government working in the north-eastern part are recommending at scientifically tempered Agro-horti-silvi pastoral system as an alternative to shifting cultivation. Despite its adverse effects on the eco-system and low productivity it still persists because it is an integral part of the socio-cultural fabric of tribal life in the northeast and is clearly linked with their religious rites and festivals e.g. Agalmaka, Miamua, Rongchugala and Ahia of Garo Hills.

Homestead Agroforestry System

A homestead or homegarden is an operational farm unit in which a number of tree species are raised along with livestock, poultry and/ or fish; mainly for the purpose of satisfying the farmers' basic needs. This farming is traditional in the Himalayan region. In a home garden multiple crops are present in a multi-tier canopy configuration. The leaf canopies of the components are arranged in such a way that they occupy different vertical layers with the tallest components having foliage tolerant to strong light and high evaporation demand and the shorter components having foliage requiring or tolerating shade and high humidity.

In Hamirpur and Kangra districts of Himachal Pradesh the homestead is constituted of Fodder trees i.e. *Celtis australis* (Khirak), *Bahaunia variegata* (Kachnar), *Grewia optiva* (Beul) in *the* upper story. The middle storey is constituted of bushes like *Adatoda vasica, Vitex negundo,* lemon and Galgal as fruit trees. During the rainy season cucurbits (vines) are grown along with Colocacia, turmeric in the field and (winter) vegetables during the winter season.

In Meghalayan homestead pisciculture is the common practice. Fishes are generally reared in a dugout ponds either in a comer or in the centre of the home garden, so that run off can be collected in it. In the low lying or plain parts paddy is generally grown. Important timber trees are *Pinus kesia* (Khasipine) *Alnus nepalensis* (Utis) *Schima wallichii* (chillowane) etc. are grown in the homestead for their requirement. In the lower storey guava, banana and vegetables are grown for domestic consumption. As per the study conducted by the ICAR Research Complex for NEH Region, Shillong; in a dugout pond of 32m × 15 m 2.2 m size there was a production of 250 kg fishes in one season.

However, wide variations in the intensity of tree cropping are found among homegardens in different locations. This can generally be attributed to the differences in socio-economic conditions of the households and their responses to externally determined changes particularly to price of the inputs and products, dependence on land and tenurial conditions among others.

Plantation Crop Combination

Plantation crops play a major role in national economics because they generate value added goods for the international market. The important plantation crops of the Himalayan region is tea. Traditionally, tea is grown on waste and marginal lands either in monoculture or in association with indigenous forest tree species. In Himachal Pradesh tea gardens in Kangra Valley are managed under the canopy *of Albizia chinensis* spp. which not only nurse the tea plants by fixing the atmospheric nitrogen in its roots but also provides shade for the development and maintenance of new tenders. The leaf litters *of Albizia* trees also add nutrients to the soil. Similarly in Assam and West Bengal the tree gardens are managed under *Alnus nepalensis* (Utis).

Large Cardamom based Agroforestry System

Large cardamom (*Amomum subulatum*) a traditional plantation crop is grown in an old typical agroforestry system. It is adapted to humid and shady environment and requires moist soil with good organic content. In natural forests it is raised in association with *Alnus nepalensis*

(Utis), a shade tree adjoining to small perennial streams (Thoras). This helps to spread water seepage through the soil during most part of the year. *Alnus nepalensis,* fast growing tree, not only produces fuel wood and timber but also fixes atmosphere nitrogen and enriches the soil by adding litter which in turn benefits the large cardamom. It also serves as a catch crop; attracting leaf feeding insect-pests to itself thereby protecting the large cardamom crop.

This silvi-horticultural system is very successfully practised in Sikkim since over a century. The area reported under large cardamom is about 14, 000 ha in Sikkim and the acreage is expected to increase phenomenally with the aid and assistance of the Cardamom Board established in Sikkim recently. In Joginder Nagar, distt. Mandi (*HP*), cultivation of large cardamom is picking up under *Celtis australis, Ficus* spp., *Alnus* spp. etc. near the perennial streams or *nalahs.*

Bamboo Groves as a Component of Agriculture Holdings (Agri-silviculture)

In Sikkim agriculture is well established as compared to other northeastern hill states. The farmer in Sikkim, by and large, adopts a mixed farming system. In all agricultural holdings along the streams and irrigation/drainage channels cultivation of bamboo (*Dendrocalamus hamiltonii* and D. *sikkimensis is* common place. Bamboo's are extensively used in building small farm houses, goat sheds, piggery enclosures, small baskets and string making. Large bamboo pipes are used as water conveyers in the farm irrigation/drainage system. Bamboo leaves serve as an excellent winter fodder for goat. Bamboo stumps/culms also protect the water channel from erosion. This system is limited only to he high rainfall areas or where the sufficient water is available to grow bamboo. A similar practice is also found in the Palampur area of Himachal Pradesh as well as in other northeastern states.

Multipurpose Trees on the Crop Land

It is a common practice in the western Himalayan region to cultivate wheat, peas, potato, cauliflower, mustard etc. during winter and maize, tomato, chillies etc. during the summer season either in monoculture or mixed on the permanent terraces prepared across the hill slopes. Fodder, fuel and timber trees viz. *Grewia optiva* (buel), *Celtis australis* (Khirak), *Bauhinia variegata* (Kachnar), *Albizzia chinensis* (Ohi), *Toona ciliata* (Toon), *Morus alba* (Toot), *Ulmus laviegata* (meryano), etc. are deliberately left or grown on the bunds of terraces. This type of practice i.e. agrisivicultural system is prevalent in sub-montane and mid hills sub-humid zone of Himachal Pradesh.

Similarly in the Agri-horticultural system which is predominantly followed in the mid hill-sub humid and high hill-temperate wet zones fruit trees are grown on terrace bunds along with agriculture crops since fuel and fodder are readily available from other sources. Agricultural crops like peas, cabbage, colocacia, turmeric and pulses are generally grown in the inter-spaces of horticultural trees such as *Malus domestica* (apple) *Prunus domestica* (plum), P. *armeniaca* (apricot) P. *persica* (peach), P. *deueis* (almond) and *Pyrus communis* (pear). Tree are uniformly spaced in the field and average density of trees is generally 5 per 100 m^2. In some cases fuel or fodder/timber trees are also retained on the field bunds and managed under agri-horti-silvicultural system. Besides providing fruits/fuelwood/fodder trees grown along bunds also restrict/.check erosion.

Private and Community Fodder Wood Blocks

Fodder tree forests in the vicinity of villages, be they on forest land, community land or private land are jointly managed by the managers to meet their fodder need for cattle during lean period i.e. winter season. In the mid and high hills of Himalayas *Quercus* spps. (Oaks) forests are jointly managed and protected for the same whereas in the foot hills or sub-mountainous region other broad leaved trees i.e. *Grewia optiva* (Buel), *Celtis australis* (Khirak), *Bauhinia variegata* (Kachnar), *Albizia chinensis* (Oei) etc. are managed so as to meet fodder and fuel requirements.

Seabuckthorn: Potential Resource

Seabuckthorn (*Hippophae* L.) an ecologically viable and ethnobotanically sustainable food crop, belonging to family Elaegnaceae, is one of the few potential resources of high mountain areas which has been reported to offer hope to give multiple benefits to poor mountain man.

Seabuckthorn, (an indigenous plant which has survived under harsh climatic conditions such as average rainfall 350-1200 mm, average temperature in the range of 10-15°C and sometimes going below -40° to -50°C), is one such specie which has immense potential for satisfying all requirements. Seabuckthorn is widely distributed along high hills, river beds, valleys and along dry mountain slopes of this region both as healthy dense stands as well as in scattered gregarious dense patches in an altitudinal range of 700 in to as high as 5500 m and above.

The people in high mountainous areas have been using seabuckthorn fruits for curing several ailments, including skin tumor diseases, digestive disorders and respiratory ailments, foot and mouth disease in cattle, besides using it extensively in agroforestry practices, though the farming community did not know of its exact potential. Seabuckthorn has also been a major source of fuelwood to local inhabitants who face scarcity of other sources of energy. Major traditional uses of Seabuckthorn in high mountainous cold and dry zone of Himachal Pradesh are fuelwood, fodder for goats and cattle (nutritious forage), fencing to protect fruit trees/nurseries/orchards, maintenance of traditional irrigation channel besides soil conservation, improving fertility/quality timber, useful for newly opened areas and subsidiary food items, medicinal products and alcoholic preparations.

In Himachal Pradesh major gene pool areas have been identified at Spiti valley, Lahaul and Kinnaur, for 4 species and 9 subspecies of genus *Hippophae* reported from different parts of the world.

The introduction, promotion and development of seabuckthorn in this area seems meaningful and will not only help in vegetation rehabilitation, but also in ecological sustenance and economic gains to local farmers of the cold and dry mountain area.

Genetic Containment of Popular Plantations

A plantation is a long, artificially-established forest, farm or estate, where crops are grown for sale, often in distant markets rather than for local on-site consumption. The term *plantation* is informal and not precisely defined. Plantations are grown on a large scale as the crops are grown for commercial purposes, not for local consumption.

Crops grown on plantations include fast-growing trees (often conifers), cotton, coffee, tobacco, sugar cane, sisal, some oil seeds (notably oil palms) and rubber trees. Farms that produce alfalfa, Lespedeza, clover, and other forage crops are usually not called plantations. The term "plantation" has usually not included large orchards (except for banana plantations), but does include the planting of trees for lumber. A plantation is always a monoculture over a large area and does not include extensive naturally occurring stands of plants that have economic value. Because of its large size, a plantation takes advantage of economies of scale. Protectionist policies and natural comparative advantage have contributed to determining where plantations have been located.

Among the earliest examples of plantations were the latifundia of the Roman Empire, which produced large quantities of wine and olive oil for export. Plantation agriculture grew rapidly with the increase in international trade and the development of a worldwide economy that followed the expansion of European colonial empires. Like every economic activity, it has changed over time. Earlier forms of plantation agriculture were associated with large disparities of wealth and income, foreign ownership and political influence, and exploitative social systems such as indentured labour and slavery. The history of the environmental, social and economic issues relating to plantation agriculture are covered in articles that focus on those subjects.

Forestry

Industrial Plantations: Industrial plantations are established to produce a high volume of wood in a short period of time for each society. Plantations are grown by state forestry authorities (for example, the Forestry Commission in Britain) and/or the paper and wood industries and other private landowners (such as Weyerhaeuser, Rayonier and Plum Creek Timber in the United States, Asia Pulp & Paper in Indonesia). Christmas trees are often grown on plantations as well. In southern and southeastern Asia, teak plantations have recently replaced the natural forest.

Industrial plantations are actively managed for the commercial production of forest products. Industrial plantations are usually large-scale. Individual blocks are usually even-aged and often consist of just one or two species. These species can be exotic or indigenous. The plants used for the plantation are often genetically altered for desired traits such as growth and resistance to pests and diseases in general and specific traits, for example in the case of timber species, volumic wood production and stem straightness. Forest genetic resources are the basis for genetic alteration. Selected individuals grown in seed orchards are a good source for seeds to develop adequate planting material.

Wood production on a tree plantation is generally higher than that of natural forests. While forests managed for wood production commonly yield between 1 and 3 cubic metres per hectare per year, plantations of fast-growing species commonly yield between 20 and 30 cubic metres or more per hectare annually; a Grand Fir plantation at Craigvinean in Scotland has a growth rate of 34 cubic metres per hectare per year (Aldhous & Low 1974), and Monterey Pine plantations in southern Australia can yield up to 40 cubic metres per hectare per year (Everard & Fourt 1974). In 2000, while plantations accounted for 5% of global forest, it is estimated that they supplied about 35% of the world's roundwood.

Growth Cycle:

- In the first year, the ground is prepared usually by the combination of burning, herbicide spraying, and/or cultivation and then saplings are planted by human crew or by machine.

The saplings are usually obtained in bulk from industrial nurseries, which may specialize in selective breeding in order to produce fast growing disease- and pest-resistant strains.

- In the first few years until the canopy closes, the saplings are looked after, and may be dusted or sprayed with fertilizers or pesticides until established.
- After the canopy closes, with the tree crowns touching each other, the plantation is becoming dense and crowded, and tree growth is slowing due to competition. This stage is termed 'pole stage'. When competition becomes too intense (for pine trees, when the live crown is less than a third of the tree's total height), it is time to thin out the section. There are several methods for thinning, but where topography permits, the most popular is 'row-thinning', where every third or fourth or fifth row of trees is removed, usually with a harvester.

  Many trees are removed, leaving regular clear lanes through the section so that the remaining trees have room to expand again. The removed trees are delimbed, forwarded to the forest road, loaded onto trucks, and sent to a mill. A typical pole stage plantation tree is 7–30 cm in diametre at breast height (dbh). Such trees are sometimes not suitable for timber, but are used as pulp for paper and particleboard, and as chips for oriented strand board.
- As the trees grow and become dense and crowded again, the thinning process is repeated. Depending on growth rate and species, trees at this age may be large enough for timber milling; if not, they are again used as pulp and chips.
- Around year 10-60 the plantation is now mature and (in economic terms) is falling off the back side of its growth curve. That is to say, it is passing the point of maximum wood growth per hectare per year, and so is ready for the final harvest. All remaining trees are felled, delimbed, and taken to be processed.
- The ground is cleared, and the cycle is repeated.

Some plantation trees, such as pines and eucalyptus, can be at high risk of fire damage because their leaf oils and resins are flammable to the point of a tree being explosive under some conditions.

Conversely, an afflicted plantation can in some cases be cleared of pest species cheaply through the use of a prescribed burn, which kills all lesser plants but does not significantly harm the mature trees.

Criticism of Industrial Plantations

Figure: *Bushfires pose a high risk to Eucalyptus plantations.*

In contrast to a naturally regenerated forest, plantations are typically grown as even-aged monocultures, primarily for timber production.

Plantations are usually near- or total monocultures. That is, the same species of tree is planted across a given area, whereas a natural forest would contain a far more diverse range of tree species.

Plantations may include tree species that would not naturally occur in the area. They may include unconventional types such as hybrids, and genetically modified trees may be used sometime in the future. Since the primary interest in plantations is to produce wood or pulp, the types of trees found in plantations are those that are best-suited to industrial applications. For example, pine, spruce and eucalyptus are widely planted far beyond their natural range because of their fast growth rate, tolerance of rich or degraded agricultural land and potential to produce large volumes of raw material for industrial use.

Plantations are always young forests in ecological terms. Typically, trees grown in plantations are harvested after 10 to 60 years, rarely up to 120 years. This means that the forests produced by plantations do not contain the type of growth, soil or wildlife typical of old-growth natural forest ecosystems. Most conspicuous is the absence of decaying dead wood, a crucial component of natural forest ecosystems.

In the 1970s, Brazil began to establish high-yield, intensively managed, short rotation plantations. These types of plantations are sometimes called fast-wood plantations or fibre farms and often managed on a short-rotation basis, as little as 5 to 15 years. They are becoming more widespread in South America, Asia and other areas. The environmental and social impacts of this type of plantation has caused them to become controversial. In Indonesia, for example, large multi-national pulp companies have harvested large areas of natural forest without regard for regeneration. From 1980 to 2000, about 50% of the 1.4 million hectares of pulpwood plantations in Indonesia have been established on what was formerly natural forest land.

The replacement of natural forest with tree plantations has also caused social problems. In some countries, again, notably Indonesia, conversions of natural forest are made with little regard for rights of the local people. Plantations established purely for the production of fibre provide a much narrower range of services than the original natural forest for the local people. India has sought to limit this damage by limiting the amount of land owned by one entity and, as a result, smaller plantations are owned by local farmers who then sell the wood to larger companies. Some large environmental organizations are critical of these high-yield plantations and are running an anti-plantation campaign, notably the Rainforest Action Network and Greenpeace.

Farm or Home Plantations

Farm or home plantations are typically established for the production of timber and fire wood for home use and sometimes for sale. Management may be less intensive than with Industrial plantations. In time, this type of plantation can become difficult to distinguish from naturally regenerated forest. Teak and bamboo plantations in India have given good results and an alternative crop solution to farmers of central India, where conventional farming was popular. But due to rising input costs of farming many farmers have done teak and bamboo plantations which require very little water (only during first two years). Teak and bamboo have legal protection from theft. Bamboo, once planted, gives output for 50 years till flowering occurs. Teak requires 20 years to grow to full maturity and fetch returns.

Environmental Plantations

These may be established for watershed or soil protection. They are established for erosion control, landslide stabilization and windbreaks. Such plantations are established to foster native species and promote forest regeneration on degraded lands as a tool of environmental restoration.

Ecological Impact

Probably the single most important factor a plantation has on the local environment is the site where the plantation is established. If natural forest is cleared for a planted forest then a reduction in biodiversity and loss of habitat will likely result. In some cases, their establishment may involve draining wetlands to replace mixed hardwoods that formerly predominated with pine species. If a plantation is established on abandoned agricultural land, or highly degraded land, it can result in an increase in both habitat and biodiversity. A planted forest can be profitably established on lands that will not support agriculture or suffer from lack of natural regeneration.

The tree species used in a plantation is also an important factor. Where non-native varieties or species are grown, few of the native fauna are adapted to exploit these and further biodiversity loss occurs. However, even non-native tree species may serve as corridors for wildlife and act as a buffer for native forest, reducing edge effect.

Once a plantation is established, how it is managed becomes the important environmental factor. The single most important factor of management is the rotation period. Plantations harvested on longer rotation periods (30 years or more) can provide similar benefits to a naturally regenerated forest managed for wood production, on a similar rotation. This is especially true if native species are used.

In the case of exotic species, the habitat can be improved significantly if the impact is mitigated by measures such as leaving blocks of native species in the plantation, or retaining corridors of natural forest. In Brazil, similar measures are required by government regulations.

Plantations and Natural Forest Loss

Many forestry experts claim that the establishment of plantations will reduce or eliminate the need to exploit natural forest for wood production. In principle this is true because due to the high productivity of plantations less land is needed.

Many point to the example of New Zealand, where 19% of the forest area provides 99% of the supply of industrial round wood. It has been estimated that the world's demand for fibre could be met by just 5% of the world forest (Sedjo & Botkin 1997). However in practice, plantations are replacing natural forest, for example in Indonesia.

According to the FAO, about 7% of the natural closed forest being lost in the tropics is land being converted to plantations. The remaining 93% of the loss is land being converted to agriculture and other uses. Worldwide, an estimated 15% of plantations in tropical countries are established on closed canopy natural forest.

In the Kyoto Protocol, there are proposals encouraging the use of plantations to reduce carbon dioxide levels (though this idea is being challenged by some groups on the grounds that the sequestered CO_2 is eventually released after harvest).

Other Types of Plantation

Crops may be called plantation crops because of their association with a specific type of farming economy. Most of these involve a large landowner, raising crops with economic value rather than for subsistence, with a number of employees carrying out the work. Often it referred to crops newly introduced to a region.

In past times it has been associated with slavery, indentured labour, and other economic models of high inequity. However, arable and dairy farming are both usually (but not always) excluded from such definitions. A comparable economic structure in antiquity was the *latifundia* that produced commercial quantities of olive oil or wine, for export. One plantation crop is bananas and there are others as well.

High Value Food Crops

Plantings of a number of trees or shrubs grown for food or beverage, including tea, coffee, and cacao are generally called plantations. Some spice and high value crops grown from permanent perennial stock, such as black pepper, may also be so called. When the holding belongs to a single individual, that person may be called a *planter*.

Sugar

Sugar plantations were highly valued in the Caribbean by the British and French colonists in the 17th and 18th centuries and the use of sugar in Europe rose during this period. Sugarcane is still an important crop in Cuba. Sugar plantations also arose in countries such as Barbados and Cuba because of the natural endowments that they had. These natural endowments included soil that was conducive to growing sugar and a high marginal product of labour realized through the increasing number of slaves.

Rubber

Plantings of para rubber, the tree *Hevea brasiliensis*, are usually called plantations.

Oil Palm

Oil palm agriculture is rapidly expanding across wet tropical regions, and is usually developed at plantation scale.

Orchards

Fruit orchards are sometimes considered to be plantations.

Arable Crops

These include tobacco, sugarcane, pineapple, and cotton, especially in historical usage. Before the rise of cotton in the American South, indigo and rice were also sometimes called plantation crops.

Fishing Plantations in Newfoundland and Labrador

When Newfoundland was colonized by England in 1610, the original colonists were called "Planters" and their fishing rooms were known as "fishing plantations". These terms were used well into the 20th century.

The following three plantations are maintained by the Government of Newfoundland and Labrador as provincial heritage sites:

- Sea-Forest Plantation was a 17th-century fishing plantation established at Cuper's Cove (present-day Cupids) under a royal charter issued by King James I.
- Mockbeggar Plantation is an 18th-century fishing plantation at Bonavista.
- Pool Plantation a 17th-century fishing plantation maintained by Sir David Kirke and his heirs at Ferryland. The plantation was destroyed by French invaders in 1696.

Other fishing plantations:

- Bristol's Hope Plantation, a 17th-century fishing plantation established at Harbour Grace, created by the Bristol Society of Merchant-Adventurers.
- Benger Plantation, an 18th-century fishing plantation maintained by James Benger and his heirs at Ferryland. It was built on the site of Georgia plantation.
- Piggeon's Plantation, an 18th-century fishing plantation maintained by Ellias Piggeon at Ferryland.

Plantations in the Antebellum American South

In the American South, antebellum plantations were centred on a "plantation house", the residence of the owner, where important business was conducted. Slavery and plantations had different characteristics in different regions of the South. As the Upper South of the Chesapeake

Bay Colony developed first, historians of the antebellum South defined planters as those who held 20 or more slaves. Major planters held many more, especially in the Deep South as it developed. The majority of slaveholders held 10 or fewer slaves, often just a few to labour domestically. By the late 18th century, most planters in the Upper South had switched from exclusive tobacco cultivation to mixed crop production, both because tobacco had exhausted the soil and because of changing markets. The shift away from tobacco meant they had slaves in excess of the number needed for labour, and they began to sell them in the internal slave trade.

There was a variety of domestic architecture on plantations. The largest and wealthiest planter families, for instance, those with estates fronting on the James River in Virginia, constructed mansions in brick and Georgian style, e.g. Shirley Plantation. Common or smaller planters in the late 18th and 19th century had more modest wood frame buildings, such as Southall Plantation in Charles City County.

In the Low Country of South Carolina, by contrast, even before the American Revolution, planters holding large rice and cotton plantations in South Carolina typically owned hundreds of slaves. In Charleston and Savannah, the elite also held numerous slaves to work as household servants. The 19th-century development of the Deep South for cotton cultivation depended on large plantations with much more acreage than was typical of the Chesapeake Bay area, and for labour, planters held hundreds of slaves.

Until December 1865 slavery was legal in parts of the United States. Most slaves were employed in agriculture, and "planter" was a term commonly used to describe a farmer with many slaves.

The term "planter" has no universally accepted definition but academic historians have defined it to identify the elite class, "a landowning farmer of substantial means." In the "Black Belt" counties of Alabama and Mississippi, the terms ""planter" and "farmer" were often synonymous. Robert Fogel and Stanley Engerman define large planters as owning over 50 slaves, and medium planters as owning between 16 and 50 slaves. In his study of Black Belt counties in Alabama, Jonathan Wiener defines planters by ownership of real property, rather than of slaves. A planter, for Wiener, owned at least $10,000 worth of real estate in 1850 and $32,000 worth in 1860, equivalent to about the top 8 percent of landowners. In his study of southwest Georgia, Lee Formwalt also defines planters in size of land holdings rather than slaves. Formwalt's planters are in the top 4.5 percent of landowners, translating into real estate worth $6,000 or more in 1850, $24,000 or more in 1860, and $11,000 or more in 1870. In his study of Harrison County, Texas, Randolph B. Campbell classifies large planters as owners of 20 slaves, and small planters as owners of between ten and 19 slaves. In Chicot and Phillips Counties, Arkansas, Carl H. Moneyhon defines large planters as owners of twenty or more slaves, and six hundred or more acres.

Genetically Modified Tree

A genetically modified tree (GMt, GM tree, genetically engineered tree, GE tree or transgenic tree) is a tree whose DNA has been modified using genetic engineering techniques. In most cases the aim is to introduce a novel trait to the plant which does not occur naturally within the species. Examples include resistance to certain pests, diseases, environmental conditions, and herbicide tolerance, or the alteration of lignin levels in order to reduce pulping costs.

Genetically modified forest trees are not yet approved ("deregulated") for commercial use, with the exception of insect-resistant poplar trees in China. Several genetically modified forest tree species are undergoing field trials for deregulation, much of the research is being carried out by the pulp and paper industry, primarily with the intention of increasing the productivity of existing tree stock. Certain genetically modified orchard tree species have been deregulated for commercial use in the United States including the papaya and plum. The development, testing and use of GM trees remains at an early stage in comparison to GM crops.

Research

Research into genetically modified trees has been ongoing since 1988. Concerns surrounding the biosafety implications of releasing genetically modified trees into the wild have held back regulatory approval of GM forest trees. This concern is exemplified in the Convention on Biological Diversity's stance:

The Conference of the Parties, Recognising the uncertainties related to the potential environmental and socio-economic impacts, including long term and trans-boundary impacts, of genetically modified trees on global forest biological diversity, as well as on the livelihoods of indigenous and local communities, and given the absence of reliable data and of capacity in some countries to undertake risk assessments and to evaluate those potential impacts, recommends parties to take a precautionary approach when addressing the issue of genetically modified trees.

A precondition for further commercialization of GM forest trees is likely to be their complete sterility. Plantation trees remain phenotypically similar to their wild cousins in that most are the product of no more than three generations of artificial selection, therefore, the risk of transgene escape by pollination with compatible wild species is high. One of the most credible science based concerns with GM trees is their potential for wide dispersal of seed and pollen. The fact that pine pollen travels long distances is well established, moving up to 3,000 kilometres from its source. Additionally, many tree species reproduce for a long time before being harvested. In combination these factors have led some to believe that GM trees are worthy of special environmental considerations over GM crops. Ensuring sterility for GM trees has proven elusive, but efforts are being made, and it is predicted that complete containment may be possible by 2020.

Proposed Uses

GM trees under experimental development have been modified with traits intended to provide benefit to industry, foresters or consumers. Due to high regulatory and research costs, the majority of genetically modified trees in silviculture consist of plantation trees, such as eucalyptus, poplar, and pine.

Lignin Alteration

Several companies in the pulp and paper industry are interested in utilizing GM technology to alter the lignin content of plantation trees. It is estimated that reducing lignin in plantation trees by genetic modification could reduce pulping costs by up to $15 per cubic metre. Lignin removal from wood fibres conventionally relies on costly and environmentally hazardous chemicals. By developing low-lignin GM trees it is hoped that pulping and bleaching processes will require fewer inputs, therefore, mills supplied by low-lignin GM trees may have a reduced impact on their surrounding ecosystems and communities. However, it is argued that reductions in lignin may compromise the structural integrity of the plant, thereby making it more susceptible to pathogens and disease, which could necessitate pesticide use exceeding that on traditional plantations.

Frost Tolerance

Genetic modification can allow trees to cope with abiotic stresses such that their geographic range is broadened. Freeze-tolerant GM eucalyptus trees for use in southern US plantations are currently being tested in open air sites with such an objective in mind. ArborGen, a tree biotechnology company and joint venture of pulp and paper firms Rubicon (New Zealand), MeadWestvaco (US) and International Paper (US) is leading this research. Until now the cultivation of eucalyptus has only been possible on the southern tip of Florida, freeze-tolerance would substantially extend the cultivation range northwards.

Accelerated Growth

In Brazil, field trials of fast growing GM eucalyptus are currently underway, they are set to conclude in 2015-2016 with commercialization to result. FuturaGene, a biotechnology company owned by Suzano, a Brazilian pulp and paper company, has been leading this research. Stanley Hirsch, chief executive of FuturaGene has stated: “Our trees grow faster and thicker. We are ahead of everyone. We have shown we can increase the yields and growth rates of trees more than anything grown by traditional breeding.” The company is looking to reduce harvest cycles from 7 to 5.5 years with 20-30% more mass than conventional eucalyptus. There is concern that such objectives may further exacerbate the negative impacts of plantation forestry. Increased water and soil nutrient demand from faster growing species may lead to irrecoverable losses in site productivity and further impinge upon neighbouring communities and ecosystems.

Disease Resistance

Ecologically motivated research into genetic modification is underway. There are ongoing schemes that aim to foster disease resistance in trees such as the American chestnut and the English elm for the purpose of their reintroduction to the wild. Specific diseases have reduced the populations of these emblematic species to the extent that they are mostly lost in the wild. Genetic modification is being pursued concurrently with traditional breeding techniques in an attempt to endow these species with disease resistance.

Current Uses

Poplars in China: In 2002 China’s State Forestry Administration approved GM poplar trees for commercial use. Subsequently 1.4 million Bt (insecticide) producing GM poplars were planted in China. They were planted both for their wood and as part of China’s ‘Green Wall’ project, which aims to impede desertification. Reports indicate that the GM poplars have spread beyond the area of original planting and that contamination of native poplars with the Bt gene is occurring. There is concern with these developments, particularly because the pesticide producing trait may impart a positive selective advantage on the poplar, allowing it a high level of invasiveness.

The Utility of Popular and Willow for Restoring

Populus is a genus of 25–35 species of deciduous flowering plants in the family Salicaceae, native to most of the Northern Hemisphere. English names variously applied to different species include poplar aspen, and cottonwood. In the September 2006 issue of *Science Magazine*, the Joint Genome Institute announced that the Western Balsam Poplar (*P. trichocarpa*) was the first tree for which its full DNA code had been determined by DNA sequencing.

Description

The genus has a large genetic diversity, and can grow from anywhere between 15–50 m (49–164 ft) tall, with trunks of up to 2.5 m (8 ft 2 in) diametre.

The bark on young trees is smooth, white to greenish or dark grey, often with conspicuous lenticels; on old trees it remains smooth in some species, but becomes rough and deeply fissured in others. The shoots are stout, with (unlike in the related willows) the terminal bud present. The leaves are spirally arranged, and vary in shape from triangular to circular or (rarely) lobed, and with a long petiole; in species in the sections *Populus* and *Aigeiros*, the petioles are laterally flattened, so that breezes easily cause the leaves to wobble back and forth, giving the whole tree a "twinkling" appearance in a breeze. Leaf size is very variable even on a single tree, typically with small leaves on side shoots, and very large leaves on strong-growing lead shoots. The leaves often turn bright gold to yellow before they fall during autumn.

The flowers are mostly dioecious (rarely monoecious) and appear in early spring before the leaves. They are borne in long, drooping, sessile or pedunculate catkins produced from buds formed in the axils of the leaves of the previous year. The flowers are each seated in a cup-shaped disk which is borne on the base of a scale which is itself attached to the rachis of the catkin. The scales are obovate, lobed and fringed, membranous, hairy or smooth, usually caducous. The male flowers are without calyx or corolla, and comprise a group of 4–60 stamens inserted on a disk; filaments short, pale yellow; anthers oblong, purple or red, introrse, two-celled; cells opening longitudinally. The female flower also has no calyx or corolla, and comprises a single-celled ovary seated in a cup-shaped disk. The style is short, with 2–4 stigmas, variously lobed, and numerous ovules. Pollination is by wind, with the female catkins lengthening considerably between pollination and maturity. The fruit is a two to four-valved dehiscent capsule, green to reddish-brown, mature in mid summer, containing numerous minute light brown seeds surrounded by tufts of long, soft, white hairs which aid wind dispersal.

Ecology

Poplars of the cottonwood section are often wetlands or riparian trees. The aspens are among the most important boreal broadleaf trees.

Poplars and aspens are important food plants for the larvae of a large number of Lepidoptera species. *Pleurotus populinus*, the aspen oyster mushroom, is found exclusively on dead wood of *Populus* trees in North America.

Classification

Figure: *Group of four poplars on a hill through the seasons, April, September, October, February (Germany)*

The genus *Populus* has traditionally been divided into six sections on the basis of leaf and flower characters; this classification is followed below. Recent genetic studies have largely supported this, confirming some previously suspected reticulate evolution due to past hybridisation and introgression events between the groups.

Some species (noted below) had differing relationships indicated by their nuclear DNA (paternally inherited) and chloroplast DNA sequences (maternally inherited), a clear indication of likely hybrid origin. Hybridisation continues to be common in the genus, with several hybrids between species in different sections known.

Selected Species

- *Populus* section *Populus* – aspens and White Poplar. Circumpolar subarctic and cool temperate, and mountains farther south (White Poplar warm temperate)
  - o *Populus tremula* – aspen, common aspen, Eurasian aspen, European aspen, quaking aspen (Europe, northern Asia)
  - o *Populus adenopoda* – Chinese Aspen (Eastern Asia)

- *Populus alba* – White Poplar (Southern Europe to central Asia)
  - *Populus* × *canescens* (*P. alba* × *P. tremula*) – Grey Poplar
  - *Populus spp. X*– Pacific Albus (North America)
- *Populus davidiana* – Korean Aspen (Eastern Asia)
- *Populus grandidentata* – Bigtooth Aspen (Eastern North America)
- *Populus sieboldii* – Japanese Aspen (Eastern Asia)
- *Populus tremuloides* – Quaking Aspen or Trembling Aspen (North America)

• *Populus* section *Aigeiros* – black poplars, some of the cottonwoods. North America, Europe, western Asia; temperate

- *Populus deltoides* – Eastern Cottonwood (Eastern North America)
- *Populus fremontii* – Fremont Cottonwood (Western North America)
- *Populus nigra* – Black Poplar (Europe) Placed here by nuclear DNA; cpDNA places in sect. *Populus*. (Including *Populus afghanica*.)
  - *Populus* × *canadensis* (*P. deltoides* × *P. nigra*) – Hybrid Black Poplar
  - *Populus* × *inopina* (*P. nigra* × *P. fremontii*) – Hybrid Black Poplar

Figure: *Leaves of* Populus lasiocarpa

- *Populus* section *Tacamahaca* – balsam poplars. North America, Asia; cool temperate
  - o *Populus angustifolia* – Willow-leaved Poplar or Narrowleaf Cottonwood (Central North America)
  - o *Populus balsamifera* – Ontario Balsam Poplar (Northern North America) (= *P. candicans, P. tacamahaca*)
  - o *Populus cathayana* – (Northeast Asia)
  - o *Populus koreana* J.Rehnder – Korean Poplar (Northeast Asia)
  - o *Populus laurifolia* – Laurel-leaf Poplar (Central Asia)
  - o *Populus maximowiczii* A.Henry – Maximowicz' Poplar, Japanese Poplar (Northeast Asia)
  - o *Populus simonii* – Simon's Poplar (Northeast Asia)
  - o *Populus suaveolens* Fischer – Mongolian Poplar (Northeast Asia)
  - o *Populus szechuanica* – Sichuan Poplar (Northeast Asia) Placed here by nuclear DNA; cpDNA places in sect. *Aigeiros*.
  - o *Populus trichocarpa* – Western Balsam Poplar or Black Cottonwood (Western North America)
  - o *Populus tristis* (Northeast Asia) Placed here by nuclear DNA; cpDNA places in sect. *Aigeiros*.
  - o *Populus ussuriensis* – Ussuri Poplar (Northeast Asia)
  - o *Populus yunnanensis* – Yunnan Poplar (East Asia)
- *Populus* section *Leucoides* – necklace poplars or bigleaf poplars. Eastern North America, eastern Asia; warm temperate
  - o *Populus heterophylla* – Swamp Cottonwood (Southeastern North America)
  - o *Populus lasiocarpa* – Chinese Necklace Poplar (Eastern Asia)
  - o *Populus wilsonii* – Wilson's Poplar (Eastern Asia)
- *Populus* section *Turanga* – subtropical poplars. Southwest Asia, east Africa; subtropical to tropical
  - o *Populus euphratica* – Euphrates Poplar (North Africa, Southwest and Central Asia)
  - o *Populus ilicifolia* – Tana River Poplar (East Africa)
- *Populus* section *Abaso* – Mexican poplars. Mexico; subtropical to tropical
  - o *Populus guzmanantlensis* (Mexico)
  - o *Populus mexicana* – Mexico Poplar (Mexico)

Cultivation

Figure: *A fastigiate Black Poplar cultivar of the Plantierensis Group, in Hungary*

Many poplars are grown as ornamental trees, with numerous cultivars used. They have the advantage of growing very big, very fast. Almost all poplars take root readily from cuttings or where broken branches lie on the ground.

Trees with fastigiate (erect, columnar) branching are particularly popular, and are widely grown across Europe and southwest Asia. However, like willows, poplars have very vigorous and invasive root systems stretching up to 40&mbsp;m from the trees; planting close to houses or ceramic water pipes may result in damaged foundations and cracked walls and pipes due to their search for moisture. A simple, reproducible, high frequency micropropagation protocol in Eastern Cottonwood *Populus deltoides* has been reported by Yadav et al. 2009.

The Poplar in India

In India, the poplar is grown commercially by farmers, mainly in the Punjab region. Popular poplar varieties are as follows:

1. G48 (grown in the plains of Punjab, Haryana, UP)
2. w22 (grown in mountainous regions, i.e. Himachal Pradesh, Pathankot, Jammu)

The poplar is grown from "kalam" (cuttings), harvested annually in January and February, and is commercially available up to 15 November. The most common use of poplar is in plywood. Yamuna Nagar in state of Haryana has a large plywood industry reliant upon poplar. It is graded according in sizes known as "over" (over 24 inches in girth), "under" (18–24 inches), and "sokta" (less than 18 inches).

Uses

Although the wood from *Populus* is known as poplar wood, a common high-quality hardwood "poplar" with a greenish colour is actually from an unrelated genus *Liriodendron*. *Populus* wood is a lighter, more porous material.

In modern society poplar is not readily associated with many uses beyond biomass. This poor reputation is undeserved, as its flexibility and close grain give it a balance of properties that have made it highly desirable for a number of applications (similar to those for willow) since antiquity. Notably the Greeks and Etruscans made shields of poplar, and Pliny also recommended poplar for this purpose. Poplar continued to be used for shield construction through the Middle Ages and was renowned for a durability similar to that of oak, but at a substantial reduction in weight.

Manufacturing

- In many areas, fast-growing hybrid poplars are grown on plantations for pulpwood
- Poplar is widely used for the manufacture of paper.
- It is also sold as inexpensive hardwood timber, used for pallets and cheap plywood; more specialised uses include matches and the boxes in which camembert cheese is sold.
- Poplar wood is also widely used in the snowboard industry for the snowboard core, because it has exceptional flexibility, and is sometimes used in the bodies of electric guitars and drums.
- Poplar wood, particularly when seasoned, makes a good hearth for a bow drill.
- Due to its high tannic acid content, the bark has been used in Europe for tanning leather.
- Poplar wood can be used to produce chopsticks.

Energy

There is interest in using poplar as an energy crop for biomass or biofuel, in energy forestry systems, particularly in light of its high energy-in / energy-out ratio, large carbon mitigation potential and fast growth. In the United Kingdom, poplar (as with fellow energy crop willow) is typically grown in a short rotation coppice system for two to five years (with single or multiple stems), then harvested and burned - the yield of some varieties can be as high as 12 oven dry tonnes every year

Land Management

Lombardy Poplars are used as a windbreak around agricultural fields to protect against wind erosion.

Agriculture

Logs from the poplar provide a growing medium for Shiitake mushrooms.

Populus Sect. Aigeiros

Populus section *Aigeiros* is a section of three species in the genus *Populus*, the poplars. Commonly known as cottonwoods, the species are native to North America, Aigeiros, and western Asia. In the past, as many as six species were recognized, but recent trends have been to accept just three species, treating the others as subspecies of *P. deltoides*.

They are large, deciduous trees 20–45 m tall, distinguished by thick, deeply fissured bark and triangular-based to diamond-shaped leaves that are green on both sides (without the whitish wax on the undersides of balsam poplar leaves) and without any obvious balsam scent in spring. An important feature of the leaves is the petiole, which is flattened sideways so that the leaves have a particular type of movement in the wind.

Male and female flowers are in separate catkins, appearing before the leaves in spring. The seeds are borne on cottony structures that allow them to be blown long distances in the air before settling to ground.

The cottonwoods are exceptionally tolerant of flooding, erosion, and flood deposits filling around the trunk.

Although each of the three cottonwood species has a different leaf pattern, they all have the same general diamond leaf shape.

Populus deltoides (eastern cottonwood) is one of the largest North American hardwood trees, although the wood is rather soft. It is a riparian zone tree. It occurs throughout the eastern United States and in extreme southern Canada. The leaves are alternate and simple, with coarsely toothed (crenate/serrate) edges, and subcordate at the base. The leaf shape is roughly triangular, hence the species name, *deltoides*.

In the typical subspecies *P. d. deltoides* (Vermont south to northern Florida and west to about Michigan), the leaves are broad and triangular, 7–15 cm across at the base. Further west (Minnesota south to eastern Texas), the subspecies *P. d. monilifera* (plains cottonwood; syn. *P. sargentii*) has somewhat narrower leaves, 5–10 cm wide at the base. This is also the state tree of Nebraska, Wyoming, and Kansas. In West Texas, New Mexico, and Colorado, the subspecies *P. d. wislizeni* (Rio Grande cottonwood; syn. *P. wislizeni*) occurs.

Populus fremontii (Fremont cottonwood) occurs in California east to Utah and Arizona and south into northwest Mexico; it is similar to the eastern cottonwood, differing mainly in the leaves having fewer, larger serrations on the edge, and small differences in the flower and seed pod structure. Some consider *P. fremontii* to be a subspecies of *P. deltoides*.

Populus nigra (black poplar), native to Europe and western Asia, is distinct in its much smaller leaves, 5–11 cm across, with a more rhombic (diamond) shape.

Cultivation and Uses

Cottonwoods are widely grown for timber production along wet river banks, where their exceptional growth rate provides a large crop of wood within just 10–30 years. The wood is coarse and of fairly low value, used for pallet boxes, shipping crates, and similar purposes where a cheap but strong enough wood is suitable. They are also widely grown as screens and shelterbelts. Many of the cottonwoods grown commercially are the hybrid of eastern cottonwood and black poplar, *Populus × canadensis* (hybrid black poplar or Carolina poplar).

Cottonwood bark is often a favourite medium for artisans. The bark, which is usually harvested in the fall after a tree's death, is generally very soft and easy to carve.

Cottonwood is one of the poorest wood fuels; it does not dry well, and rots quickly. It splits poorly, because it is very fibrous. It produces a low level of energy per unit of volume of wood.

Cottonwoods serve as food for the caterpillars of several Lepidoptera (list of Lepidoptera that feed on poplars).

Aspen

Aspen is a common name for certain tree species; some, but not all, are classified by botanists in the section *Populus*, of the poplar genus.

Species

These species are called aspens:

- *Populus adenopoda* – Chinese Aspen (China, south of *P. tremula*)
- *Populus davidiana* – Korean Aspen (Eastern Asia)
- *Populus grandidentata* – Bigtooth Aspen (eastern North America, south of *P. tremuloides*)
- *Populus sieboldii* – Japanese Aspen (Japan)
- *Populus tremula* – Eurasian Aspen (northern Europe and Asia)
- *Populus tremuloides* – Quaking Aspen or Trembling Aspen (northern and western North America)

Habitat and Longevity

The aspens are all native to cold regions with cool summers, in the north of the Northern Hemisphere, extending south at high altitudes in the mountains. They are all medium-sized deciduous trees reaching 15–30 m (49–98 ft) tall.

All of the aspens typically grow in large clonal colonies, derived from a single seedling, and spread by means of root suckers; new stems in the colony may appear at up to 30–40 m (98–131 ft) from the parent tree. Each individual tree can live for 40–150 years above ground, but the root system of the colony is long-lived. In some cases, this is for thousands of years, sending up new trunks as the older trunks die off above ground. For this reason, it is considered to be an indicator of ancient woodlands. One such colony in Utah, given the nickname of "Pando", is estimated to be 80,000 years old, making it possibly the oldest living colony of aspens. Some aspen colonies become very large with time, spreading about 1 m (3.3 ft) per year, eventually covering many hectares. They are able to survive forest fires, because the roots are below the heat of the fire, with new sprouts growing after the fire burns out.

Aspens do not thrive in the shade, and it is difficult for seedlings to grow in an already mature aspen stand. Fire indirectly benefits aspen trees, since it allows the saplings to flourish in open sunlight in the burned landscape. Lately, aspens have an increased popularity in forestry, mostly because of their fast growth rate and ability to regenerate from sprouts, making the reforestation after harvesting much cheaper, since no planting or sowing is required.

In contrast with many trees, aspen bark is base-rich, meaning aspens are important hosts for bryophytes and act as food plants for the larvae of butterfly (Lepidoptera) species.

Young aspen bark is an important seasonal forage for the European hare and other animals in early spring. Aspen is also a tree of choice of the European beaver.

Uses

Aspen wood is white and soft, but fairly strong, and has low flammability. It has a number of uses, notably for making matches and paper where its low flammability makes it safer to use than most other woods. Shredded aspen wood is used for packing and stuffing, sometimes called excelsior (wood wool). It is also a popular animal bedding, since it lacks the phenols associated with pine and juniper, which are thought to cause respiratory system ailments in some animals. Heat-treated aspen is a popular material for the interiors of a sauna. While standing trees sometimes tend to rot from the heart outward, the dry timber weathers very well, becoming silvery-grey and resistant to rotting and warping, and has traditionally been used for rural construction in the northwestern regions of Russia (especially for roofing, in the form of thin slats).

3

Economic Potential for Sustainable Harvest of Medicinal/Economic Plants

Plants are the source of a large proportion of medicines. It has recently been estimated that there are about 400,000 species of higher or vascular plants. While not all of these are equally useful, studies suggest that somewhere between a quarter and a third of all species have been used for medicinal purposes by indigenous societies. Medicinal plants have been identified and used throughout human history. Toxic plants even have use in pharmaceutical development. Angiosperms (flowering plants) were the original source of most plant medicines. Some herbs and spices come from flowering plants.

Deans and Svoboda (1990) suggested that there are over 20,000 medicinal plants; Torkelson (1995) lists over 4,000 scientific names and more than 28,000 common names of medicinal plants; and Duke (2002) provides detailed information on over 800 of the most important medicinal plant species. Despite these large numbers, relatively few species are cultivated commercially on a large scale, as is the case, for example, with ginseng, echinacea, and ginkgo. For medicinal plants, becoming commercially significant is not easy. The odds are about a thousand to one. A species needs to be very special in order to join the relatively elite club of important commercial plants. Fortunately, with common sense and enough background information, the odds of profitably growing medicinal plants that have not previously been raised in a given region are considerably higher than a thousand to one.

While this discussion focuses on the economic wisdom of expanding production of a specific class of crop (medicinal crops) in a specific province of Canada, it is very important to remember that this is an example of a general problem. The phrases "alternative crops," "replacement crops," "substitute crops" and "new crops" all relate to attempts to improve the profitability of farming by growing crops that have not been grown, at least recently, in a give locality. Most such crops are "minor," "speciality," "niche," "non-traditional," or "under exploited," at least initially in the region in question.

There have been many analyses of the exercise of new crop introduction. It is disappointing to learn that no one has yet found a reliable means of choosing new crops that will be successful. Nevertheless, by examining some aspects of new crop introduction, it is at least possible to make more informed choices and to avoid some problems. Except in cases of monopoly, new offerings in the marketplace strongly tend to follow a characteristic life cycle of profitability, and it is extremely important that those investing in any new crop understand this.

The figure provided here illustrates the trend in a hypothetical fashion. Phase 1 is the period of investment in research and development required to bring the new product to the point of profitability; most new products do not survive this period. During phase 2, the market for the product expands to the point of saturation, becoming more profitable.

Phase 3 is a stable period of profitability, at the end of which decline occurs (phase 4). The most common cause of the decline in profitability is copy-cat competition and the more profitable and popular a given product becomes, the faster competition develops.

A frequent consequence of such competition is over-saturation of the market, the generation of surpluses, and business failures of those who remained dependent on the sale of the once profitable item. It really doesn't matter how good or important the product is - even the cereals upon which the world depends have become questionable investments because of surplus production. (The profitability life-cycle diagram is pessimistic in suggesting that every crop will eventually become unprofitable, but there is an escape from this fate: research leading to new cultivars and new uses.)

Echinacea is presently at the end of phase 3, and has entered phase 4. For those who are already growing it, there are still profits to be made, but it is a bad idea for new investors to start growing echinacea. Because of its formidable reputation, however, there are still many who want to jump on the echinacea bandwagon. For every crop, there is an ideal number of producers. Too little production means that insufficient material can be placed on the market to sustain interest. Too much production means that prices are depressed and profits decrease. In addition to the need for there to be a balance between production and the marketplace, there are also minimum needs for specialties associated with post-production aspects.

For many new crops, limitations of processing and/or storage facilities are critical bottlenecks. Often, storage and processing capacity exists, but is remote from the sites of production, so that transportation is the problem. Ideally, there is balance among producers, processors, and marketers, as has generally developed for major, established crops and crop products.

Almost inevitably, the development of new crops is accompanied by growing pains, and very often there is a period of several years during which some, who were enthusiastic initially, abandon the enterprise. Frequently a key to success of a new crop is the backing of an influential manufacturer or retailer, which can result in sufficient market exposure for the merits of the product to become widely appreciated. Certainly supportive producer and marketing organizations are helpful, if not indeed indispensable.

Economic Botany

Economic botany is the commercial exploitation of plants by people. Economic botany contributes significantly to anthropology, biology, conservation, botany, and other fields of science. This link between botany and anthropology explores the ways humans use plants for food, shelter, medicines, textiles, and much more.

Botany itself came about through medicine and the development of herbal remedies. Thus at its advent, botany was economic as well as systematic. As plants became useful for herbals and curatives, their economic value increased. An early set of instructions drawn up by a cosmographer of Charles the fifth instructed explorers to "determine what are the items of sustenance of the land and which onse are generally used, whether fruits or seeds, and all manner of spices, drugs, or whatever other scents, and find out the time in which one can reproduce the trees, plants, herbs, and fruits that these parts offer, and if the natives use them for medicines, as we do."

Teosinte and rice are two examples of plants modified so that their economic values would increase.

Teosinte

The teosintes are grasses of the genus *Zea*. Native Americans bred and selected teosinte for the traits we see in corn today (large ears, multiple rows of kernels). The first ears of maize were very short, with only 8 rows of kernels. Modern corn is the result of several thousand generations of selective breeding. Modern corn is incapable of reproducing without human help; the kernels will stay firmly attached to the cob and rot. This doesn't represent a useful adaptation for the species, but is excellent for harvesting and transporting corn.

Rice

Rice was first domesticated approximately 5,000 years ago, in Southeast Asia. Rice and American wild rice are believed to have been domesticated separately. Rice variants have been adapted to the tropics where they provide a grain staple, but rice can be grown almost anywhere. The introduction of dwarf rice variants made several rice-producing countries self-sufficient. Rice is suited to countries with high rainfall.

Economically Important Food Plants

Plants that humans use for food are of high economic importance. Research into food plants generally involves increasing the size of the edible plant organ in question, or increasing the areas where the plant can be grown, and less frequently, finding new crop species. Results of such research are often published in the journal *Economic Botany*. The New Zealand-based Plant & Food Research publishes its own journal on cultivar development and sustainable production systems for high quality produce, and the design and development of new and novel functional foods.

Florida Oranges

Citrus has been a major commercial product in Florida since the 19th century. Florida produces over 70% of the U.S. citrus supply. The colour of oranges is not related to ripening, but is a serious component for sales. The orange colour only develops in areas with cool nighttime temperatures. In tropical climates, growers often expose the fruit to ethylene, to promote the loss of chlorophyll and expose the beta-carotenes (the orange colour).

North American Apples

Apples are not native to North America, but today the North American continent boasts the greatest diversity of apples in the world. Part of this is due to "Johnny Appleseed," real name John Chapman. Chapman spent 48 years travelling all along the American northwest spreading apple seeds and planting trees. While apples come in literally thousands of varieties, the majority of the apple market is based on three: Red Delicious, Golden Delicious, and Granny Smith. The Red

Delicious is the ideal apple, for marketing purposes, for the apple industry. Its large, intensely red, and instantly recognizable.

Economically Valuable Medicinal Plants

Medical research in the U.S. alone has a budget of $95 billion. A large portion of that money is spent on research into plants and plant extracts. Several key medical discoveries have been made by studying plants and the compounds they produce, to see the effect they have on humans.

Ephedrine

The Ephedra plant is the natural source of ephedrine, the plants principle alkaloid. Ephedrine is actually a very interesting case of economic botany in medicine. While it has been taken medicinally in the past, ephedrine can be highly toxic. Because of this fact, medical researchers studied the compound and produced pseudoephedrine. This is the medicine you can buy over the counter, in Sudafed and other decongestants. Ephedrine imitates epinephrine in its effect on the human body. Originally developed by the plant as a herbivore deterrent, this compound, studied and refined by researchers, now helps fix allergy symptoms nationwide every year.

Echinacea

One of many herbal remedies out there, Echinacea represents a sizable industry. Many people take echinacea for cold and flu-like symptoms, but studies show that the plant has had mixed success fighting these viruses. However, those same studies show the plant possibly being useful for the treatment of upper respiratory infections. NCCAM is currently studying echinacea for the treatment of upper respiratory infections as well as its effect on the immune system.

Ornamental Plants

Ornamental plants can be found in almost any store, and many people have at least one in their home. However, ornamental plants are not limited to houseplants. Landscaping agencies make heavy use of ornamental plants, usually with an accompanying high cost. Trees, shrubs, flowers, and grasses, all of these are planted by professional landscaping agencies regularly, with a large economic effect.

Why New Crops May Not Be Good Investments

Too much optimism is as bad as too much pessimism. In the following, a general case against new crops is made, simply to emphasize the potential hazards.

Farmers are conservative, that is, they put great faith in what experience has taught is successful. This is common sense, because changes are normally risky, and so much can go wrong in farming that taking unnecessary risks is foolish. Farmers are comfortable with crops that they have grown for years, are adapted to their soils and machinery, and for which there are established markets. For very similar reasons, processors, marketers and others in the agri-food industry are reluctant to embrace change for change's sake.

For example, it is much easier to conduct business when local, national and international markets are well known. New crops or commodities made from them must penetrate the existing marketing or industrial infrastructure, and this is usually beyond the control of farmers contemplating the new crops.

There are several key potential negative financial consequences that can result from getting into new crops:

- the economic gamble may be lost: new crops chosen may turn out to be less profitable than the old crops
- long-term lack of profit: successful new crops often require a lengthy period of development and commercialization before any profit is realized
- new crops may simply displace old profitable crops, with no net profit
- copy-cats may adapt the crop to their circumstances and out-compete those who have established it, limiting the period of profitability

Seeking new crops is simply an economic decision that the risk of continuing to rely exclusively

on traditional crops is probably not as wise as trying new ones.

Questions to Ask Regarding Development of a New Crop

1. What demand is being met?
2. Who will grow it?
3. Who will process it?
4. Who will sell it?
5. Who will buy it?
6. Who will consume it?
7. What commodities are being replaced?
8. On whose toes will you be stepping?

9. Who should pay for the development costs?
10. Can intellectual property be protected?

Special Investment Problems with Medicinal Plants

Most medicinal species that might seem worth developing as new crops are undomesticated. These often have a limited yield, mature irregularly, are very variable, aren't suited to current planting and harvesting machinery, or present many other problems such as:

- seeds are often difficult to obtain
- expertise and knowledge is often in very short supply (and often are regarded as trade secrets)
- plant breeding may be necessary
- management techniques will have to be developed
- markets will have to be located or developed

From time to time, popular medicinal herbs are found to contain toxic constituents, and they lose their popularity, or are even prohibited by government. The cyclical popularity of the herbal market is a major problem, and an investment can be greatly lowered in value in a short period. For a novice, without farming experience, getting into medicinal plant production is a dangerous decision. For an experienced farmer who has not previously dealt with medicinal plants, it would be wise to not bet the family farm on medicinal plants alone.

In the last 3 years, the medicinal plant over-the-counter industry in North America has experienced reduced sales, particularly because of bad publicity arising from quality issues. The industry is currently attempting to overcome these problems. It might seem that this is not an ideal moment for investment. Conversely, however, just as sophisticated stockbrokers buy when they think the market has bottomed out, this may be exactly the ideal time for investment!

Woodland Cultivation

Numerous medicinal species occur in woodlands as slow-growing perennials. Many of these plants have evolved toxic chemicals to defend themselves against insects and other life forms and, as is so often the case, these chemicals have proven to be useful as medicinal agents.

This is an excellent area for investment in research. It should be possible to develop sustainable ways of growing medicinal species in at least some classes of woodlands. There are obvious economic advantages, since investment would be much less than in strictly agricultural production systems.

Organic Production

Depending on crop, organic production may add appreciable value to a crop. In some cases, organic crops may be demanded by a wholesaler or processor. Certainly organic production offers North Americans a great advantage over much of the Third World medicinal plant supply, whose certification as organic is sometimes not reliable.

Sustainable Harvesting and Management of Medicinal Plants

People have used medicinal plants (MPs) in health care since the time of earliest human evolution. In many countries, MPs still remain the major source of medication for a wide range of ailments. Therapeutic effects of MPs are associated with their chemical peculiarities, which are in reality components of the defense strategies of plants. Only 5 to 15% of the approximately 270,000 species of higher plants occurring in the globe have been systematically investigated, and plants remain a rich source of novel bioactive compounds. The potential value of medicinal compounds derived from plants has been proposed as a tangible benefit of biodiversity and therefore a basis for promoting its conservation.

The role of MPs is particularly important in the Himalayan region, where a large proportion of the rural population depends on wild MP resources to meet their health care needs. Besides their importance in health care, MPs have high socio-cultural, symbolic and economic values, providing income and employment to millions of people living in the region. In the Himalaya, MPs are highly valued in different folk healing systems and in scholarly systems of traditional medicine, such as Ayurveda, Unani, Chinese and Tibetan. In addition, Himalayan medicinal plants and their products are also popular in other traditional medical systems (such as Siddha, homoeopathic) as well as in modern allopathic medicines. Plants used in traditional medicine are important sources of novel bio-molecules, with application for the manufacture of pharmaceuticals and cosmeceuticals.

Besides the incorporation of medicinal herbs into health foods, dietary supplements, herbal teas, cosmetics, massage oils, fragrances, and dying and colouring agents has dramatically increased the international demand in medicinal plants. The global market for herbal medicines has been estimated to be worth USD 40-60 billion, annually growing at the rate of 7-10%. The Himalayan region is one of the major repositories of high value MPs in Asia concerned by international trade. India and China alone consume hundreds of MP species from the Himalaya. The annual demand for raw materials of MPs in India, for example, is estimated at 24 million metric tons, which is increasing at the rate of 20% per annum. Similarly, the traditional Chinese medicine market has been valued at US$ 4830 million, which is projected to reach over 1.21 billion by the year 2010.

The recent trend of large-scale commercial extraction of MPs, to meet the growing demand of herbal industries, has increased the probability of over-exploitation of many valuable species. The pace, depth, and magnitude of human pressures in the Himalaya are known to have exerted ecological stresses not only on the individuals and the populations of MPs but also on the life support ecosystems.

This has lead to the degradation of diversity, quality and availability of many valuable species on the one hand and their habitats and ecosystems on the other. Knowledge of the sustainability of the use of such plant resources is thus urgently needed. In the Himalaya, there is a general lack of precise scientific knowledge regarding both the biological and socio-cultural aspects of the resource base and potential for sustainable harvest.

Collection, Storage, and Use

In recent years there has been a tremendous surge of interest in herbal or botanic medicine. In consequence, a wide variety of ready-prepared natural remedies have found their way onto the shelves of herb suppliers, health-food stores, and even some chemists. While manufacturers are to be commended for presenting the public with an alternative to chemicals, it's important to remember that a lot of remedies can be made at home. While naturally treating what ails you may seem like hocus pocus, many wonderful cures are as common as what's found in your herb garden.

Collecting Medicinal Plants

No matter where you live, you will always be able to find plants that can be cultivated for their medicinal values. Of course, exactly what plants are available to you depends largely on your own geographical location. Where I live, one can find an almost endless variety of useful herbs, "weeds", and shrubs. Even in quite heavily populated towns it is normally possible to find an inexhaustible supply of dandelion, groundsel, chickweed, coltsfoot, dock, plaintain, and bindweed.

When collecting plants—especially those to be used for medicinal purposes—there are several golden rules to follow. Stick to them, and you can be sure that the plants you pick will be of the finest quality.

Rule 1: Correctly Identify the Plant

Some herbs are almost indistinguishable from others that have totally different properties. Rosebay Willow-herb *(Chamaenerion angustifolium)* may, to the untrained eye, look suspiciously like Purple Loosestrife *(Lythrum salicaria)*, and two members of the Figwort family Scrophulariaceae—Yellow Lute *(Odontites lutea)* and Common Cow-wheat *(Melampyrum pratense)*—may also be easily confused by someone who is gingerly taking his or her first herb-hunting expedition in the fields.

Always take several illustrated pocket books with you when you are looking for herbs that you are not overly familiar with.

Rule 2 : Never Pick within One Mile of a Highway

Some plants such as the Raspberry, Blackberry, Lesser Plaintain, Ground Ivy, and Self-Heal have a curious affinity for the lead thrown out by car-exhaust fumes. Plants picked by a busy roadside may contain up to 200 times their natural level of lead.

Rule 3 : Always Pick from the Correct Area

It is not by coincidence that you find a field covered with a golden carpet of Coltsfoot or a meadow dotted profusely with clumps of Plantain. If an herb is growing prolifically in a particular area, you can be sure that the soil is rich in the minerals and organic acids or alkalines that promote vigorous and healthy growth. Try to pick your herbs from areas such as these.

Rule 4 : Carefully Choose the Time You Pick

Pick by mid-morning, when most of the dew has evaporated from the foliage (dew-laden herbs tend to develop mold soon after picking).

Forest Farming

Forest farming is the cultivation of high-value speciality crops under a forest canopy that is intentionally modified or maintained to provide shade levels and habitat that favour growth and enhance production levels. Forest farming encompasses a range of cultivated systems from introducing plants into the understory of a timber stand to modifying forest stands to enhance the marketability and sustainable production of existing plants.

Forest farming is a type of agroforestry practice characterized by the "four I's": intentional, integrated, intensive and interactive. Agroforestry is a land management system that combines trees with crops or livestock, or both, on the same piece of land. It focuses on increasing benefits to the landowner as well as maintaining forest integrity and environmental health. The practice involves cultivating non-timber forest products or niche crops, some of which, such as ginseng or shiitake mushrooms, can have high market value.

Non-timber forest products (NTFPs) are plants, parts of plants, fungi, and other biological materials harvested from within and on the edges of natural, manipulated, or disturbed forests. Examples of crops are ginseng, shiitake mushrooms, decorative ferns, and pine straw. Products typically fit into the following categories: edible, medicinal and dietary supplements, floral or decorative, or speciality wood-based products.

History

Forest farming, though not always by that name, is practised around the world. For centuries, humans have relied on fruits, nuts, seeds, parts of foliage and pods from trees and shrubs in the forests to feed themselves and their livestock. Over time, certain species have been selected for cultivation near homes or livestock to provide food or medicine. For example, in the southern United States, mulberry trees are used as a feedstock for pigs and often cultivated near pig quarters.

In 1929, J. Russell Smith, Emeritus Professor of Economic Geography at Columbia University, published "Tree Crops – A Permanent Agriculture" which stated that crop-yielding trees could provide useful substitutes for cereals in animal feeding programmes, as well as conserve environmental health. Toyohiko Kagawa read and was heavily influenced by Smith's publication and began experimental cultivation under trees in Japan during the 1930s. Through forest farming, or three-dimensional forestry, Kagawa addressed problems of soil erosion by persuading many of Japan's upland farmers to plant fodder trees to conserve soil, supply food and feed animals. He combined extensive plantings of walnut trees, harvested the nuts and fed them to the pigs, then sold the pigs as a source of income.

When the walnut trees matured, they were sold for timber and more trees were planted so that there was a continuous cycle of economic cropping that provided both short-term and long-term income to the small landowner. The success of these trials prompted similar research in other countries. Unfortunately, World War II disrupted communication and slowed advances in forest farming. In the mid-1950s research resumed in places such as southern Africa. Kagawa was also an inspiration to Robert Hart pioneered forest gardening in temperate climates in the sixties in Shropshire, England.

In earlier years, livestock were often considered part of the forest farming system. Now they are typically excluded and agroforestry systems that integrate trees, forages and livestock are referred to as silvopastures. Because forest farming combines ecological stability of natural forests and productive agriculture systems, it is considered to have great potential for regenerating soils, restoring ground water supplies, controlling floods and droughts and cultivating marginal lands. In addition to these benefits for re-establishing productive forests on marginal lands, forest farming is way to add financial value while conserving land that is currently forested, as discussed in the methods section.

In more recent years, there has been growing interest in locally grown and organic foods throughout the United States. There has been an increase in farmer's markets and community-supported agriculture small enterprises. These have also become outlets for NTFPs. In order to stay competitive, many farmers look to add unique crops to their product line. With the quantity and quality of resources developing online that offer tutorials and educational information on how to create and maintain forest farms, forest gardens, how to cultivate specific crops such as shiitake mushrooms and how to successfully market these items, forest farming is expanding as a viable land management practice. Good places to look for research-based resources are the USDA National Agroforestry Centre's publication section, the Centre for Agroforestry at the University of Missouri, the Cornell Cooperative Extension, the Non-Timber Forest Products website by The Virginia Tech Department of Wood Science and Forest Products, the USDA Forest Service Southern Research Station and the Top of the Ozarks RC&D in Missouri and the collaborative Forest Farming community of practice on eXtension.org, the online presence of the Cooperative Extension System of the US Land Grant Universities.

Principles

Forest farming principles constitute an ecological approach to forest management. Forest resources are judiciously used while biodiversity and wildlife habitat are conserved. Forest farms have the potential to restore ecological balance to fragmented second growth forests through intentional manipulation to create the desired forest ecosystem.

In some instances, the intentional introduction of species for botanicals, medicinals, food or decorative products is accomplished using existing forests. The tree cover, soil type, water supply, land form and other site characteristics determine what species will thrive. Developing an understanding of species/site relationships as well as understanding the site limitations is necessary to utilize these resources for production needs, while conserving adequate resources for the long-term health of the forest.

Apart from the environmental benefits, forest farming can increase the economic value of forest property and provide short- and long-term benefits to the landowner. Forest farming provides economic return from intact forest ecosystems, but timber sales can remain part of the long-term management strategy.

Methods

Forest farming methods may include: Intensive, yet careful thinning of overstocked, suppressed tree stands; multiple integrated entries to accomplish thinning so that systemic shock is minimized; and interactive management to maintain a cross-section of healthy trees and shrubs of all ages and species. Physical disturbance to the surrounding area should be minimized. The following are forest farming techniques described in the Agroforestry Training Manual produced by the Centre for Agroforestry at the University of Missouri.

Level of Management Required

Forest gardening is the most intensive of forest farming methods. In addition to thinning the overstory, this method involves clearing the understory of undesirable vegetation and other practices that are closely related to agronomy (tillage, fertilization, weeding, and control of disease and insects and wildlife management). Due to input levels, this method often produces lower valued products compared to other methods. Forest gardens take advantage of the vertical levels of light availability and space under the forest canopy so that more than one crop can be grown at once if desired.

Wild-simulated seeks to maintain a natural growing environment, yet enriches local NTFP populations to create an abundant renewable supply of the products. Minimal disturbance and natural growing conditions ensure products will be similar in appearance and quality of those harvested from the wild. Rather than till, practitioners often rake leaves to expose soil, sow seed directly onto the ground, and then cover with leaves again. Since this method produces NTFPs that closely resemble wild plants; they often command a higher price than NTFPs produced using the forest gardening method.

Forest tending involves adjusting tree crown density to manipulate light levels that favour natural reproduction of desirable NTFPs. This low intensity management approach does not involve supplemental planting to increase populations of desired NTFPs.

Wildcrafting is the harvesting of naturally growing NTFPs. It is not considered a forest farming practice since there is no human involvement in the plant's establishment and maintenance. However, wildcrafters often take steps to protect NTFPs with future harvests in mind. It becomes agroforestry once forest thinnings, or other inputs, are applied to sustain or maintain plant populations that might otherwise succumb to successional changes in the forest.

The most important difference between forest farming and wildcrafting is that forest farming intentionally produces NTFPS, whereas wildcrafting seeks and gathers from naturally growing NTFPs.

Production Considerations

Forest farming can be a small business opportunity for landowners and requires careful planning, including a business and marketing plan. Learning how to market the NTFPs on the Internet is an option, but may entail higher shipping costs. Landowners should consider all options for selling their products including, farmer's markets or restaurants that focus on locally grown ingredients.

The development phase should include a forest management plan that states the landowner's objectives and a resource inventory. Start-up costs should be analyzed as specific equipment may be necessary to harvest or process the product, whereas other crops require minimal initial investment. Local incentives for sustainable forest management, as well as regulations and policies should be explored. The Convention on International Trade in Endangered Species of Wild Fauna and Flora (CITES) regulates international trade of certain plant (American ginseng and goldenseal) and animal species. To be legally exported, regulated plants must be harvested and records kept according to CITES rules and restrictions. Many states also have harvesting regulations for certain native plants that are searchable online. Another good source to start with on information is the Medicinal Plants at Risk 2008 report by the Centre for Biological Diversity in the U.S.

Examples of Crops

Medicinal herbs:

- Ginseng (*Panax quinquefolius*)
- Black Cohosh (*Actaea racemosa*)
- Goldenseal (*Hydrastis canadensis*)
- Bloodroot (*Sanguinaria canadensis*)
- Pacific yew (*Taxus brevifolia*)

- Mayapple (*Podophyllum peltatum*)
- Saw palmetto (*Serenoa repens*)
- American Pokeweeed (*Phytolacca americana*)

Nuts:

- Black walnut (*Juglans nigra*)
- Hazelnut (*Corylus avellana*)
- Shagbark hickory (*Carya ovata*)
- Beechnut (*Fagus sylvatica*)

Fruit:

- Pawpaw (*Asimina triloba*)
- Currants (*Ribes spp*)
- Elderberry (*Sambucas spp*)
- Serviceberry (*Amelanchier spp*)
- Blackberry (*Rubus spp*)
- Huckleberry (*Gaylussacia brachycera*)

Other food crops:

- Ramps (wild leeks) (*Allium tricoccum*)
- Syrups (maple)
- Honey
- Mushrooms
- Other edible roots

Other products: (mulch, decoratives, crafts, dyes)

- Pine straw
- Willow twigs
- Vines
- Beargrass (*Xerophyllum tenax*)
- Ferns
- Pine cones
- Moss

Native ornamentals:

- Rhododendron (*Rhododendron catawbiense*)
- Highbush cranberry (*Viburnum trilobum*)
- Flowering dogwood (*Cornus florida*)

Approach and Practicable Techniques of Bioengineering

Biological engineering or bioengineering (including biological systems engineering) is the application of concepts and methods of biology (and secondarily of physics, chemistry, mathematics, and computer science) to solve real-world problems related to the life sciences or the application thereof, using engineering's own analytical and synthetic methodologies and also its traditional sensitivity to the cost and practicality of the solution(s) arrived at. In this context, while traditional engineering applies physical and mathematical sciences to analyze, design and manufacture inanimate tools, structures and processes, biological engineering uses primarily the rapidly developing body of knowledge known as molecular biology to study and advance applications of living organisms.

An especially important application is the analysis and cost-effective solution of problems related to human health, but the field is much more general than that. For example, biomimetics is a branch of biological engineering which strives to find ways in which the structures and functions of living organisms can be used as models for the design and engineering of materials and machines. Systems biology, on the other hand, seeks to utilize the engineer's familiarity with complex artificial systems, and perhaps the concepts used in "reverse engineering", to facilitate the difficult process of recognition of the structure, function, and precise method of operation of complex biological systems.

The differentiation between biological engineering and biomedical engineering can be unclear, as many universities loosely use the terms "bioengineering" and "biomedical engineering" interchangeably. Biomedical engineers are specifically focused on applying biological and other sciences towards medical innovations, whereas biological engineers are focused principally on applying engineering principles to biology - but not necessarily for medical uses. Hence neither "biological" engineering nor "biomedical" engineering is wholly contained within the other, as there can be "non-biological" products for *medical* needs as well as "biological" products for *non-medical* needs (the latter including notably biosystems engineering).

History

Biological engineering is a science-based discipline founded upon the biological sciences in the same way that chemical engineering, electrical engineering, and mechanical engineering can be based upon chemistry, electricity and magnetism, and classical mechanics, respectively.

Biological engineering can be differentiated from its roots of pure biology or other engineering fields. Biological studies often follow a reductionist approach in viewing a system on its smallest possible scale which naturally leads towards tools such as functional genomics. Engineering approaches, using classical design perspectives, are constructionist, building new devices, approaches, and technologies from component concepts. Biological engineering utilizes both kinds of methods in concert, relying on reductionist approaches to identify, understand, and organize the fundamental units which are then integrated to generate something new. In addition, because it is an engineering discipline, biological engineering is fundamentally concerned with not just the basic science, but its practical application of the scientific knowledge is to solve real-world problems in a cost-effective way. Although engineered biological systems have been used to manipulate information, construct materials, process chemicals, produce energy, provide food, and help maintain or enhance human health and our environment, our ability to quickly and reliably engineer biological systems that behave as expected is at present less well developed than our mastery over mechanical and electrical systems.

ABET, the U.S.-based accreditation board for engineering B.S. programmes, makes a distinction between biomedical engineering and biological engineering, though there is much overlap. Foundational courses are often the same and include thermodynamics, fluid and mechanical dynamics, kinetics, electronics, and materials properties. According to Professor Doug Lauffenberger of MIT, biological engineering (like biotechnology) has a broader base which applies engineering principles to an enormous range of size and complexities of systems ranging from the molecular level - molecular biology, biochemistry, microbiology, pharmacology, protein chemistry, cytology, immunology, neurobiology and neuroscience (often but not always using biological substances) - to cellular and tissue-based methods (including devices and sensors), whole macroscopic organisms (plants, animals), and up increasing length scales to whole ecosystems.

The word bioengineering was coined by British scientist and broadcaster Heinz Wolff in 1954. The term bioengineering is also used to describe the use of vegetation in civil engineering construction. The term bioengineering may also be applied to environmental modifications such as surface soil protection, slope stabilisation, watercourse and shoreline protection, windbreaks, vegetation barriers including noise barriers and visual screens, and the ecological enhancement of an area. The first biological engineering programme was created at Mississippi State University in 1967, making it the first biological engineering curriculum in the United States. More recent programmes have been launched at MIT and Utah State University.

Description

Biological engineers or *bioengineers* are engineers who use the principles of biology and the tools of engineering to create usable, tangible, economically viable products. Biological engineering employs knowledge and expertise from a number of pure and applied sciences, such as mass and heat transfer, kinetics, biocatalysts, biomechanics, bioinformatics, separation and purification processes, bioreactor design, surface science, fluid mechanics, thermodynamics, and polymer science. It is used in the design of medical devices, diagnostic equipment, biocompatible materials, renewable bioenergy, ecological engineering, agricultural engineering, and other areas that improve the living standards of societies.

In general, biological engineers attempt to either mimic biological systems to create products or modify and control biological systems so that they can replace, augment, or sustain chemical and mechanical processes. Bioengineers can apply their expertise to other applications of engineering and biotechnology, including genetic modification of plants and microorganisms, bioprocess engineering, and biocatalysis. Because other engineering disciplines also address living organisms (e.g., prosthetics in bio-mechanical engineering), the term biological engineering can be applied more broadly to include agricultural engineering and biotechnology, which notably can address non-healthcare objectives as well (unlike biomedical engineering). In fact, many old agricultural engineering departments in universities over the world have rebranded themselves as agricultural and biological engineering or agricultural and biosystems engineering. Biological engineering is also called bioengineering by some colleges and biomedical engineering is called bioengineering by others, and is a rapidly developing field with fluid categorization. Depending on the institution and particular definitional boundaries employed, some major fields of bioengineering may be categorised as (note these may overlap):

- Biological systems engineering
- Biomedical engineering: biomedical technology, biomedical diagnostics, biomedical therapy, biomechanics, biomaterials.
- Genetic engineering (involving both of the above, although in different applications): synthetic biology, horizontal gene transfer.
- Bioprocess engineering: bioprocess design, biocatalysis, bioseparation, bioinformatics, bioenergy

- Cellular engineering: cell engineering, tissue engineering, metabolic engineering.
- Biomimetics: The use of knowledge gained from reverse engineering evolved living systems to solve difficult design problems in artificial systems.

Land and Farm Resources

Trees play a crucial role in almost all terrestrial ecosystems and provide a range of products and services to rural and urban people. As natural vegetation is cleared for agriculture and other types of development, the benefits that trees provide are best sustained by integrating trees into agriculturally productive landscapes — a practice known as agroforestry.

Farmers have practised agroforestry for years. Agroforestry focuses on the wide range of working trees grown on farms and in rural landscapes. Among these are fertilizer trees for land regeneration, soil health and food security; fruit trees for nutrition; fodder trees that improve smallholder livestock production; timber and fuelwood trees for shelter and energy; medicinal trees to combat disease; and trees that produce gums, resins or latex products. Many of these trees are multipurpose, providing a range of benefits.

Agroforestry provides many livelihood and environmental benefits, including:

- Enriching the asset base of poor households with farm-grown trees.
- Enhancing soil fertility and livestock productivity on farms.
- Linking poor households to markets for high-value fruits, oils, cash crops and medicines.
- Balancing improved productivity with the sustainable management of natural resources.
- Maintaining or enhancing the supply of environmental services in agricultural landscapes for water, soil health, carbon sequestration and biodiversity.

The World Agroforestry Centre's Role

The world population has now surpassed 6.5 billion. Demand for food increases as populations grow. Meanwhile, over one billion people continue to endure lives of extreme poverty. Agroforestry is uniquely suited to address the need to grow more food and biomass for fuel while sustainably managing agricultural landscapes for the critical ecosystem services they provide. It can serve as a means of curbing greenhouse gas emissions by slowing forest conversion to farmland and sequestering more carbon in trees on farms. With over three decades of work with smallholder farmers in Africa, Asia and Latin America, and strategic alliances with advanced laboratories, national research institutions, universities and non-governmental organizations, the World Agroforestry Centre is uniquely positioned to address global challenges.

To improve the livelihoods of poor smallholders and improve the sustainability and productivity of agricultural landscapes, we are:

- Broadening the range and diversity of trees that can be integrated into farming systems, especially as many produce higher income per unit of area than annual crops, require less labour and are more resilient to drought.
- Maximizing the productivity of agroforestry systems through improved tree germplasm, integrated soil fertility and the enhanced supply of high-quality tree fodder resources.
- Improving the income of poor households by facilitating their access to markets. This is also important in stabilizing land-use change in some areas, as well as increasing farmers' investment in agroforestry trees and systems.
- Working in agricultural landscapes that experience the greatest environmental stress to balance improved productivity with the sustainable management of natural resources. For example, stabilizing forest margins in Southeast Asia by converting slash-and-burn systems, and rehabilitating degraded agricultural land throughout Africa.
- Managing trees in agricultural landscapes to ensure the health of river and groundwater systems.
- Examining reward systems or other types of institutional and policy innovations (such as for carbon or water) to sustain biodiversity at the interface between smallholder agricultural landscapes and conservation areas.

Agriculture Biotechnology in the Forestry Sector

Forest Biotechnologies in Developing Countries

Forests and other wooded areas perform key economic and ecological functions. Not only do they provide goods and livelihoods but they also protect soils, regulate water and absorb carbon. Forests also shelter much of the world's biodiversity. The world has slightly less than 4 billion hectares of forests, covering about 30 percent of the world's land area. The production of wood and non-wood forest products is the primary function for 34 percent of the world's forests. More than half of all forests are used for wood and non-wood production in combination with other functions such as soil and water protection, biodiversity conservation and recreation. Only 7 percent of forests in the world are in plantations, with the balance found in natural or semi-natural, largely unmanaged and undomesticated forest stands. Planted forests are expanding, and their contribution to global industrial wood production is approaching 50 percent of the total. About 1.6 billion people rely heavily on forest resources for their livelihoods.

The forestry sector differs from the crop or livestock sectors in a number of important ways, such as that forest trees are highly heterozygous long-lived perennials with late sexual maturity and a lengthy regeneration cycle, which places high priority on retaining genetic diversity as an insurance policy against rapid change; most forest tree species have narrow regional adaptation, so the numbers of species used for planting are much higher than for food crops; forest trees serve as keystone species in dynamic ecosystems, so managing against loss translates into more than tree survival; forest trees are largely undomesticated although a few species have had some population-level improvement for one to four generations.

For management of naturally regenerated forests, DNA-based and biochemical markers are available for a growing number of tropical species. Today, findings are available to guide operational forest management plans, including in developing countries, but only for a very limited number of the hundreds of tree species that are managed in naturally regenerated tropical forests. This area of forest biotechnology continues to expand, moving from tools development into more hypothesis-driven knowledge acquisition. Such research enquiry is a powerful source of pertinent knowledge for protecting tropical forests. This research is also moving from molecular markers into genomics. Biotechnology tools such as molecular markers and genomics are therefore providing important knowledge about naturally regenerated tropical forests and important insights into the nature of the entire tropical forest ecosystems, including the relationship between the forest trees and the microbial communities with which they interact, which can influence the strategies employed for managing tropical forests.

For planted forests, although there is some overlap, the range of biotechnologies used is generally quite different from that used for naturally regenerated forests. Plantations can have different types of management systems (e.g. intensive, semi-intensive) and use different types of genetic material (e.g. wild material, genetically improved trees). Depending on the level of management intensity and the genetic material used in the planted forest, different groups of biotechnologies can be used. For simplicity, three different groups of biotechnologies can be identified according to the type of planted forests, ranging from the least sophisticated to the most advanced. A first group of biotechnologies is suitable for the least intensively managed planted forests, and includes a range of vegetative propagation methods (including micropropagation based on tissue culture), biofertilizers and genetic fingerprinting using molecular markers. It could also be complemented by conventional technologies, such as early-stage tree improvement programmes.

A second group of biotechnologies can be used for planted forests that provide industrial raw materials on a large planting scale. The single species used for plantations may be indigenous or exotic, but these plantations are intensively managed. This group of biotechnologies includes somatic embryogenesis (a tissue culture technique), molecular markers and quantitative trait locus (QTL) analyses, whole genome sequencing and functional genomics. A third and most sophisticated group of biotechnologies includes backward and reverse genomics approaches, whole-genome sequencing, low-cost vegetative propagation and genetic modification of forest trees. To date, the only report of commercial plantings with genetically modified (GM) trees is for poplar, in China. However, most tree species used in planted forests have been successfully modified at the experimental level, and traits that have been the subject of extensive research include stem shape, herbicide resistance, flowering characteristics, lignin content, insect and fungal resistance. Many developing countries currently have biosafety regulations for agricultural crops, including fruit-trees, although many others lack such frameworks and the capacity to implement them. There are no regulations, however, specific to the use of GM forest trees. Although policies and regulations adopted for agricultural crops are also likely to be used for forest trees, they present special challenges (long time frames and life spans, wild resource, major constituents of an ecosystem). Forests are not only trees, and forest ecosystems are more fragile, longer-lived and less closely controlled than crop fields. Decision-making is complicated by the fact that while agriculture is primarily viewed as a production system, forests are generally viewed as a natural system, important not only for the conservation of biodiversity but also for social and cultural values. Thus, the use of GM forest trees is viewed more as a political and environmental issue than as a technical or trade issue.

4

Forest, Deforestation and Reforestation

A forest, also referred to as a wood or the woods, is an area with a high density of trees. As with cities, depending on various cultural definitions, what is considered a forest may vary significantly in size and have different classifications according to how and of what the forest is composed. A forest is usually an area filled with trees but any tall densely packed area of vegetation may be considered a forest, even underwater vegetation such as kelp forests, or non-vegetation such as fungi, and bacteria. Tree forests cover approximately 9.4 percent of the Earth's surface (or 30 percent of total land area), though they once covered much more (about 50 percent of total land area). They function as habitats for organisms, hydrologic flow modulators, and soil conservers, constituting one of the most important aspects of the biosphere.

A typical tree forest is composed of the overstory (canopy or upper tree layer) and the understory. The understory is further subdivided into the shrub layer, herb layer, and also the moss layer and soil microbes. In some complex forests, there is also a well-defined lower tree layer. Forests are central to all human life because they provide a diverse range of resources: they store carbon, aid in regulating the planetary climate, purify water and mitigate natural hazards such as floods. Forests also contain roughly 90 percent of the world's terrestrial biodiversity.

The word "forest" comes from Middle English *forest*, from Old French *forest* (also *forès*) "forest, vast expanse covered by trees"; first introduced in English as the word for wild land set aside for hunting without the necessity in definition for the existence of trees (James 1981;Muir 2000,2008). Possibly a borrowing (probably via Frankish or Old High German) of the Medieval Latin word *foresta* "open wood", *foresta* was first used by Carolingian scribes in the Capitularies of Charlemagne to refer specifically to the king's royal hunting grounds. The term was not endemic to Romance languages (e.g. native words for "forest" in the Romance languages evolved out of the Latin word *silva* "forest, wood" (English *sylvan*); cf. Italian, Spanish, Portuguese *selva*; Romanian *silvã*; Old French *selve*); and cognates in Romance languages, such as Italian *foresta*, Spanish and Portuguese *floresta*, etc. are all ultimately borrowings of the French word.

The exact origin of Medieval Latin *foresta* is obscure. Some authorities claim the word derives from the Late Latin phrase *forestam silvam*, meaning "the outer wood"; others claim the term is a latinisation of the Frankish word **forhist* "forest, wooded country", assimilated to *forestam silvam* (a common practice among Frankish scribes). Frankish **forhist* is attested by Old High German *forst* "forest", Middle Low German *vorst* "forest", Old English *fyrhþ* "forest, woodland, game preserve, hunting ground" (English *frith*), and Old Norse *fýri* "coniferous forest", all of which derive from Proto-Germanic **furhísa-*, **furhíþija-* "a fir-wood, coniferous forest", from Proto-Indo-European **perkʷu-* "a coniferous or mountain forest, wooded height".

Uses of the word "forest" in English to denote any uninhabited area of non-enclosure are now considered archaic. The word was introduced by the Norman rulers of England as a legal term (appearing in Latin texts like the Magna Carta) denoting an uncultivated area legally set aside for hunting by feudal nobility. These hunting forests were not necessarily wooded much, if at all. However, as hunting forests did often include considerable areas of woodland, the word "forest" eventually came to mean wooded land more generally. By the start of the fourteenth century the word appeared in English texts, indicating all three senses: the most common one, the legal term and the archaic usage.

Other terms used to mean "an area with a high density of trees" are *wood, woodland, wold, weald, holt, frith* and *firth*. Unlike *forest*, these are all derived from Old English and were not borrowed from another language. Some classifications now reserve the term *woodland* for an area with more open space between trees and distinguish among woodlands, *open forests*, and *closed forests* based on crown cover.

Distribution

Forests can be found in all regions capable of sustaining tree growth, at altitudes up to the tree line, except where natural fire frequency or other disturbance is too high, or where the environment has been altered by human activity.

The latitudes 10° north and south of the Equator are mostly covered in tropical rainforest, and the latitudes between 53°N and 67°N have boreal forest. As a general rule, forests dominated by angiosperms (*broadleaf forests*) are more species-rich than those dominated by gymnosperms (*conifer*, *montane*, or *needleleaf forests*), although exceptions exist.

Forests sometimes contain many tree species only within a small area (as in tropical rain and temperate deciduous forests), or relatively few species over large areas (e.g., taiga and arid montane coniferous forests). Forests are often home to many animal and plant species, and biomass per unit area is high compared to other vegetation communities. Much of this biomass occurs below ground in the root systems and as partially decomposed plant detritus. The woody component of a forest contains lignin, which is relatively slow to decompose compared with other organic materials such as cellulose or carbohydrate.

Forests are differentiated from woodlands by the extent of canopy coverage: in a forest, the branches and the foliage of separate trees often meet or interlock, although there can be gaps of varying sizes within an area referred to as forest. A woodland has a more continuously open canopy, with trees spaced farther apart, which allows more sunlight to penetrate to the ground between them.

Among the major forested biomes are:

- rain forest (tropical and temperate)
- taiga
- temperate hardwood forest
- tropical dry forest

Types of Forests

Forests can be classified in different ways and to different degrees of specificity. One such way is in terms of the "biome" in which they exist, combined with leaf longevity of the dominant species (whether they are evergreen or deciduous). Another distinction is whether the forests are composed predominantly of broadleaf trees, coniferous (needle-leaved) trees, or mixed.

- Boreal forests occupy the subarctic zone and are generally evergreen and coniferous.
- Temperate zones support both broadleaf deciduous forests (*e.g.*, temperate deciduous forest) and evergreen coniferous forests

(*e.g.*, temperate coniferous forests and temperate rainforests). Warm temperate zones support broadleaf evergreen forests, including laurel forests.

- Tropical and subtropical forests include tropical and subtropical moist forests, tropical and subtropical dry forests, and tropical and subtropical coniferous forests.
- Physiognomy classifies forests based on their overall physical structure or developmental stage (e.g. old growth vs. second growth).
- Forests can also be classified more specifically based on the climate and the dominant tree species present, resulting in numerous different forest types (e.g., ponderosa pine/Douglas-fir forest).

A number of global forest classification systems have been proposed, but none has gained universal acceptance. UNEP-WCMC's forest category classification system is a simplification of other more complex systems (e.g. UNESCO's forest and woodland 'subformations'). This system divides the world's forests into 26 major types, which reflect climatic zones as well as the principal types of trees. These 26 major types can be reclassified into 6 broader categories: temperate needleleaf; temperate broadleaf and mixed; tropical moist; tropical dry; sparse trees and parkland; and forest plantations. Each category is described as a separate section below.

Temperate Needleleaf

Temperate needleleaf forests mostly occupy the higher latitude regions of the northern hemisphere, as well as high altitude zones and some warm temperate areas, especially on nutrient-poor or otherwise unfavourable soils. These forests are composed entirely, or nearly so, of coniferous species (Coniferophyta). In the Northern Hemisphere pines *Pinus*, spruces *Picea*, larches *Larix*, firs *Abies*, Douglas firs *Pseudotsuga* and hemlocks *Tsuga*, make up the canopy, but other taxa are also important. In the Southern Hemisphere, most coniferous trees (members of the Araucariaceae and Podocarpaceae) occur in mixtures with broadleaf species, and are classed as broadleaf and mixed forests.

Temperate Broadleaf and Mixed

Temperate broadleaf and mixed forests include a substantial component of trees in the Anthophyta. They are generally characteristic of the warmer temperate latitudes, but extend to cool temperate ones, particularly in the southern hemisphere. They include such forest types as the mixed deciduous forests of the United States and their counterparts in China and Japan, the broadleaf evergreen rainforests of Japan, Chile and Tasmania, the sclerophyllous forests of Australia, central Chile, the Mediterranean and California, and the southern beech Nothofagus forests of Chile and New Zealand.

Tropical Moist

There are many different types of tropical moist forests, although most extensive are the lowland evergreen broad leaf rainforests(, for example várzea and igapó forests and the terra firma forests of the Amazon Basin; the peat swamp forests, dipterocarp forests of Southeast Asia; and the high forests of the Congo Basin. Forests located on mountains are also included in this category, divided largely into upper and lower montane formations on the basis of the variation of physiognomy corresponding to changes in altitude.

Tropical Dry

Tropical dry forests are characteristic of areas in the tropics affected by seasonal drought. The seasonality of rainfall is usually reflected in the deciduousness of the forest canopy, with most trees being leafless for several months of the year. However, under some conditions, e.g. less fertile soils or less predictable drought regimes, the proportion of evergreen species increases and the forests are characterised as "sclerophyllous". Thorn forest, a dense forest of low stature with a high frequency of thorny or spiny species, is found where drought is prolonged, and especially where grazing animals are plentiful. On very poor soils, and especially where fire is a recurrent phenomenon, woody savannas develop.

Sparse Trees and Parkland

Sparse trees and parkland are forests with open canopies of 10–30% crown cover. They occur principally in areas of transition from forested to non-forested landscapes. The two major zones in which these ecosystems occur are in the boreal region and in the seasonally dry tropics. At high latitudes, north of the main zone of boreal forest or taiga, growing conditions are not adequate to maintain a continuous closed forest cover, so tree cover is both sparse and discontinuous. This vegetation is variously called open taiga, open lichen woodland, and forest tundra. It is species-poor, has high bryophyte cover, and is frequently affected by fire.

Forest Plantations

Forest plantations, generally intended for the production of timber and pulpwood increase the total area of forest worldwide. Commonly mono-specific and/or composed of introduced tree species, these ecosystems are not generally important as habitat for native biodiversity.

However, they can be managed in ways that enhance their biodiversity protection functions and they are important providers of ecosystem services such as maintaining nutrient capital, protecting watersheds and soil structure as well as storing carbon. They may also play an important role in alleviating pressure on natural forests for timber and fuelwood production.

Forest Categories

28 forest categories are used to enable the translation of forest types from national and regional classification systems to a harmonised global one:

Temperate and Boreal Forest Types

1. Evergreen needleleaf forest – Natural forest with > 30% canopy cover, in which the canopy is predominantly (> 75%) needleleaf and evergreen.
2. Deciduous needleleaf forests – Natural forests with > 30% canopy cover, in which the canopy is predominantly (> 75%) needleleaf and deciduous.
3. Mixed broadleaf/needleleaf forest – Natural forest with > 30% canopy cover, in which the canopy is composed of a more or less even mixture of needleleaf and broadleaf crowns (between 50:50% and 25:75%).
4. Broadleaf evergreen forest – Natural forests with > 30% canopy cover, the canopy being > 75% evergreen and broadleaf.
5. Deciduous broadleaf forest – Natural forests with > 30% canopy cover, in which > 75% of the canopy is deciduous and broadleaves predominate (> 75% of canopy cover).
6. Freshwater swamp forest – Natural forests with > 30% canopy cover, composed of trees with any mixture of leaf type and seasonality, but in which the predominant environmental characteristic is a waterlogged soil.
7. Sclerophyllous dry forest – Natural forest with > 30% canopy cover, in which the canopy is mainly composed of sclerophyllous broadleaves and is > 75% evergreen.
8. Sparse trees and parkland – Natural forests in which the tree canopy cover is between 10–30%, such as in the steppe regions of the world. Trees of any type (e.g., needleleaf, broadleaf, palms).

9. Disturbed natural forest – Any forest type above that has in its interior significant areas of disturbance by people, including clearing, felling for wood extraction, anthropogenic fires, road construction, etc.
10. Exotic species plantation – Intensively managed forests with > 30% canopy cover, which have been planted by people with species not naturally occurring in that country.
11. Native species plantation – Intensively managed forests with > 30% canopy cover, which have been planted by people with species that occur naturally in that country.
12. *Unspecified forest plantation – Forest plantations showing extent only with no further information about their type, This data currently only refers to the Ukraine.
13. *Unclassified forest data – Forest data showing forest extent only with no further information about their type.

Those marked * have been created as a result of data holdings which do not specify the forest type, hence 26 categories are quoted, not 28 shown here.

Tropical Forest Types

1. Lowland evergreen broadleaf rain forest – Natural forests with > 30% canopy cover, below 1,200 m (3,937 ft) altitude that display little or no seasonality, the canopy being >75% evergreen broadleaf.
2. Lower montane forest – Natural forests with > 30% canopy cover, between 1200–1800 m altitude, with any seasonality regime and leaf type mixture.
3. Upper montane forest – Natural forests with > 30% canopy cover, above 1,800 m (5,906 ft) altitude, with any seasonality regime and leaf type mixture.
4. Freshwater swamp forest – Natural forests with > 30% canopy cover, below 1,200 m (3,937 ft) altitude, composed of trees with any mixture of leaf type and seasonality, but in which the predominant environmental characteristic is a waterlogged soil.
5. Semi-evergreen moist broadleaf forest – Natural forests with > 30% canopy cover, below 1,200 m (3,937 ft) altitude in which between 50–75% of the canopy is evergreen, > 75% are broadleaves, and the trees display seasonality of flowering and fruiting.
6. Mixed broadleaf/needleleaf forest – Natural forests with > 30% canopy cover, below 1,200 m (3,937 ft) altitude, in which the

canopy is composed of a more or less even mixture of needleleaf and broadleaf crowns (between 50:50% and 25:75%).

7. Needleleaf forest – Natural forest with > 30% canopy cover, below 1,200 m (3,937 ft) altitude, in which the canopy is predominantly (> 75%) needleleaf.
8. Mangroves – Natural forests with > 30% canopy cover, composed of species of mangrove tree, generally along coasts in or near brackish or seawater.
9. Deciduous/semi-deciduous broadleaf forest – Natural forests with > 30% canopy cover, below 1,200 m (3,937 ft) altitude in which between 50–100% of the canopy is deciduous and broadleaves predominate (> 75% of canopy cover).
10. Sclerophyllous dry forest – Natural forests with > 30% canopy cover, below 1,200 m (3,937 ft) altitude, in which the canopy is mainly composed of sclerophyllous broadleaves and is > 75% evergreen.
11. Thorn forest – Natural forests with > 30% canopy cover, below 1,200 m (3,937 ft) altitude, in which the canopy is mainly composed of deciduous trees with thorns and succulent phanerophytes with thorns may be frequent.
12. Sparse trees and parkland – Natural forests in which the tree canopy cover is between 10–30%, such as in the savannah regions of the world. Trees of any type (e.g., needleleaf, broadleaf, palms).
13. Disturbed natural forest – Any forest type above that has in its interior significant areas of disturbance by people, including clearing, felling for wood extraction, anthropogenic fires, road construction, etc.
14. Exotic species plantation – Intensively managed forests with > 30% canopy cover, which have been planted by people with species not naturally occurring in that country.
15. Native species plantation – Intensively managed forests with > 30% canopy cover, which have been planted by people with species that occur naturally in that country.

Forest Loss and Management

The scientific study of forest species and their interaction with the environment is referred to as forest ecology, while the management of forests is often referred to as forestry. Forest management has changed considerably over the last few centuries, with rapid changes from the 1980s onwards culminating in a practice now referred to as sustainable forest management. Forest ecologists concentrate on forest patterns and processes, usually with the aim of elucidating cause-and-effect relationships. Foresters who practice sustainable forest management focus on the integration of ecological, social, and economic values, often in consultation with local communities and other stakeholders.

Anthropogenic factors that can affect forests include logging, urban sprawl, human-caused forest fires, acid rain, invasive species, and the slash and burn practices of swidden agriculture or shifting cultivation. The loss and re-growth of forest leads to a distinction between two broad types of forest, primary or old-growth forest and secondary forest. There are also many natural factors that can cause changes in forests over time including forest fires, insects, diseases, weather, competition between species, etc. In 1997, the World Resources Institute recorded that only 20% of the world's original forests remained in large intact tracts of undisturbed forest. More than 75% of these intact forests lie in three countries—the Boreal forests of Russia and Canada and the rainforest of Brazil.

Canada has about 4,020,000 square kilometres (1,550,000 sq mi) of forest land. More than 90% of forest land is publicly owned and about 50% of the total forest area is allocated for harvesting. These allocated areas are managed using the principles of sustainable forest management, which includes extensive consultation with local stakeholders. About eight percent of Canada's forest is legally protected from resource development (Global Forest Watch Canada) (Natural Resources Canada). Much more forest land—about 40 percent of the total forest land base—is subject to varying degrees of protection through processes such as integrated land use planning or defined management areas such as certified forests (Natural Resources Canada).

These maps represent only virgin forest lost. Some regrowth has occurred but not to the age, size, or extent of 1620, due to population increases and food cultivation. By December 2006, over 1,237,000 square kilometres of forest land in Canada (about half the global total) had been certified as being sustainably managed (Canadian Sustainable Forestry Certification Coalition). Clearcutting, first used in the latter half of the 20th century, is less expensive, but devastating to the environment, and companies are required by law to ensure that harvested areas are adequately regenerated. Most Canadian provinces have regulations limiting the size of clearcuts, although some older clearcuts can range upwards of 110 square kilometres (27,000 acres) in size which were cut over several years. China instituted a ban on logging, beginning in 1998, due to the erosion and flooding that it caused.

In 2010, the Food and Agriculture Organization of the United Nations reported that world deforestation, mainly the conversion of tropical forests to agricultural land, had decreased over the past ten years but still continues at an alarmingly high rate in many countries. Globally, around 13 million hectares of forests were converted to other uses or lost through natural causes each year between 2000 and 2010 as compared to around 16 million hectares per year during the 1990s. The study covered 233 countries and areas. Brazil and Indonesia, which had the highest loss of forests in the 1990s, have significantly reduced their deforestation rates. In addition, ambitious tree planting programmes in countries such as China, India, the United States and Viet Nam - combined with natural expansion of forests in some regions - have added more than seven million hectares of new forests annually. As a result the net loss of forest area was reduced to 5.2 million hectares per year between 2000 and 2010, down from 8.3 million hectares annually in the 1990s.

In the United States, most forests have historically been affected by humans to some degree, though in recent years improved forestry practices has helped regulate or moderate large scale or severe impacts. However, the United States Forest Service estimates a net loss of about 2 million hectares (4,942,000 acres) between 1997 and 2020; this estimate includes conversion of forest land to other uses, including urban and suburban development, as well as afforestation and natural reversion of abandoned crop and pasture land to forest. However, in many areas of the United States, the area of forest is stable or increasing, particularly in many northern states. The opposite problem from flooding has plagued national forests, with loggers complaining that a lack of thinning and proper forest management has resulted in large forest fires.

Old-growth forest contains mainly natural patterns of biodiversity in established seral patterns, and they contain mainly species native to the region and habitat. The natural formations and processes have not been affected by humans with a frequency or intensity to change the natural structure and components of the habitat. Secondary forest contains significant elements of species which were originally from other regions or habitats. Smaller areas of woodland in cities may be managed as Urban forestry, sometimes within public parks. These are often created for human benefits; Attention Restoration Theory argues that spending time in nature reduces stress and improves health, while forest schools and kindergartens help young people to develop social as well as scientific skills in forests. These typically need to be close to where the children live, for practical logistics.

Control

Reducing Emissions

Main international organizations including the United Nations and the World Bank, have begun to develop programmes aimed at curbing deforestation. The blanket term Reducing Emissions from Deforestation and Forest Degradation (REDD) describes these sorts of programmes, which use direct monetary or other incentives to encourage developing countries to limit and/or roll back deforestation. Funding has been an issue, but at the UN Framework Convention on Climate Change (UNFCCC) Conference of the Parties-15 (COP-15) in Copenhagen in December 2009, an accord was reached with a collective commitment by developed countries for new and additional resources, including forestry and investments through international institutions, that will approach USD 30 billion for the period 2010–2012. Significant work is underway on tools for use in monitoring developing country adherence to their agreed REDD targets. These tools, which rely on remote forest monitoring using satellite imagery and other data sources, include the Centre for Global Development's FORMA (Forest Monitoring for Action) initiative and the Group on Earth Observations' Forest Carbon Tracking Portal. Methodological guidance for forest monitoring was also emphasized at COP-15. The environmental organization Avoided Deforestation Partners leads the campaign for development of REDD through funding from the U.S. government.

In evaluating implications of overall emissions reductions, countries of greatest concern are those categorized as High Forest Cover with High Rates of Deforestation (HFHD) and Low Forest Cover with High Rates of Deforestation (LFHD). Afghanistan, Benin, Botswana, Burma, Burundi, Cameroon, Chad, Ecuador, El Salvador, Ethiopia, Ghana, Guatemala, Guinea, Haiti, Honduras, Indonesia, Liberia, Malawi, Mali, Mauritania, Mongolia, Namibia, Nepal, Nicaragua, Niger, Nigeria, Pakistan, Paraguay, Philippines, Senegal, Sierra Leone, Sri Lanka, Sudan, Togo, Uganda, United Republic of Tanzania, Zimbabwe are listed as having Low Forest Cover with High Rates of Deforestation (LFHD). Brazil, Cambodia, Democratic Peoples Republic of Korea, Equatorial Guinea, Malaysia, Solomon Islands, Timor-Leste, Venezuela, Zambia are listed as High Forest Cover with High Rates of Deforestation (HFHD).

Payments for Conserving Forests

In Bolivia, deforestation in upper river basins has caused environmental problems, including soil erosion and declining water quality. An innovative project to try and remedy this situation involves landholders in upstream areas being paid by downstream water users to conserve forests. The landholders receive US$20 to conserve the trees, avoid polluting livestock practices, and enhance the biodiversity and forest carbon on their land. They also receive US$30, which purchases a beehive, to compensate for conservation for two hectares of water-sustaining forest for five years. Honey revenue per hectare of forest is US$5 per year, so within five years, the landholder has sold US$50 of honey. The project is being conducted by Fundación Natura Bolivia and Rare Conservation, with support from the Climate & Development Knowledge Network.

Farming

New methods are being developed to farm more intensively, such as high-yield hybrid crops, greenhouse, autonomous building gardens, and hydroponics. These methods are often dependent on chemical inputs to maintain necessary yields. In cyclic agriculture, cattle are grazed on farm land that is resting and rejuvenating. Cyclic agriculture actually increases the fertility of the soil. Intensive farming can also decrease soil nutrients by consuming at an accelerated rate the trace minerals needed for crop growth.The most promising approach, however, is the concept of food forests in permaculture, which consists of agroforestal systems carefully designed to mimic natural forests, with an emphasis on plant and animal species of interest for food, timber and other uses. These systems have low dependence on fossil fuels and agro-chemicals, are highly self-maintaining, highly productive, and with strong positive impact on soil and water quality, and biodiversity.

Monitoring Deforestation

There are multiple methods that are appropriate and reliable for reducing and monitoring deforestation. One method is the "visual interpretation of aerial photos or satellite imagery that is labour-intensive but does not require high-level training in computer image processing or extensive computational resources". Another method includes hot-spot analysis (that is, locations of rapid change) using expert opinion or coarse resolution satellite data to identify locations for detailed digital analysis with high resolution satellite images. Deforestation is typically assessed by quantifying the amount of area deforested, measured at the present time. From an environmental point of view, quantifying the damage and its possible consequences is a more important task, while conservation efforts are more focused on forested land protection and development of land-use alternatives to avoid continued deforestation. Deforestation rate and total area deforested, have been widely used for monitoring deforestation in many regions, including the Brazilian Amazon deforestation monitoring by INPE.

Forest Management

Efforts to stop or slow deforestation have been attempted for many centuries because it has long been known that deforestation can cause environmental damage sufficient in some cases to cause societies to collapse. In Tonga, paramount rulers developed policies designed to prevent conflicts between short-term gains from converting forest to farmland and long-term problems forest loss would cause, while during the 17th and 18th centuries in Tokugawa, Japan, the shoguns developed a highly sophisticated system of long-term planning to stop and even reverse deforestation of the preceding centuries through substituting timber by other products and more efficient use of land that had been farmed for many centuries. In 16th-century Germany, landowners also developed silviculture to deal with the problem of deforestation. However, these policies tend to be limited to environments with *good rainfall*, *no dry season* and *very young soils* (through volcanism or glaciation). This is because on older and less fertile soils trees grow too slowly for silviculture to be economic, whilst in areas with a strong dry season there is always a risk of forest fires destroying a tree crop before it matures.

In the areas where "slash-and-burn" is practised, switching to "slash-and-char" would prevent the rapid deforestation and subsequent degradation of soils. The biochar thus created, given back to the soil, is not only a durable carbon sequestration method, but it also is an extremely beneficial amendment to the soil. Mixed with biomass it brings the creation of terra preta, one of the richest soils on the planet and the only one known to regenerate itself.

Sustainable Practices

Certification, as provided by global certification systems such as Programme for the Endorsement of Forest Certification and Forest Stewardship Council, contributes to tackling deforestation by creating market demand for timber from sustainably managed forests. According to the United Nations Food and Agriculture Organization (FAO), "A major condition for the adoption of sustainable forest management is a demand for products that are produced sustainably and consumer willingness to pay for the higher costs entailed. Certification represents a shift from regulatory approaches to market incentives to promote sustainable forest management. By promoting the positive attributes of forest products from sustainably managed forests, certification focuses on the demand side of environmental conservation." Rainforest Rescue argues that the standards of organizations like FSC are too closely connected to industry interests and therefore do not guarantee environmentally and socially responsible forest management. In reality, monitoring systems are inadequate and various cases of fraud have been documented worldwide.

***Figure:** Bamboo is advocated as a more sustainable alternative for cutting down wood for fuel.*

Some nations have taken steps to help increase the amount of trees on Earth. In 1981, China created National Tree Planting Day Forest and forest coverage had now reached 16.55% of China's land mass, as against only 12% two decades ago

Using fuel from bamboo rather than wood results in cleaner burning, and since bamboo matures much faster than wood, deforestation is reduced as supply can be replenished faster.

Reforestation

Reforestation is the natural or intentional restocking of existing forests and woodlands that have been depleted, usually through deforestation. Reforestation can be used to improve the quality of human life by soaking up pollution and dust from the air, rebuild natural habitats and ecosystems, mitigate global warming since forests facilitate biosequestration of atmospheric carbon dioxide, and harvest for resources, particularly timber.

The term *reforestation* is similar to afforestation, the process of restoring and recreating areas of woodlands or forests that may have existed long ago but were deforested or otherwise removed at some point in the past. Sometimes the term *re-afforestation* is used to distinguish between the original forest cover and the later re-growth of forest to an area. Special tools, e.g. tree planting bar, are used to make planting of trees easier and faster.

Management

Reforestation of large areas can be done through the use of measuring rope (for accurate plant spacing) and dibbers, (or wheeled augers for planting the larger trees) for making the hole in which a seedling or plant can be inserted. Indigenous soil inoculants (e.g., Laccaria bicolour) can optionally be used to increase survival rates in hardy environments.

A debatable issue in managed reforestation is whether or not the succeeding forest will have the same biodiversity as the original forest. If the forest is replaced with only one species of tree and all other vegetation is prevented from growing back, a monoculture forest similar to agricultural crops would be the result. However, most reforestation involves the planting of different feedlots of seedlings taken from the area often of multiple species. Another important factor is the natural regeneration of a wide variety of plant and animal species that can occur on a clear cut. In some areas the suppression of forest fires for hundreds of years has resulted in large single aged and single species forest stands. The logging of small clear cuts and or prescribed burning, actually increases the biodiversity in these areas by creating a greater variety of tree stand ages and species.

For harvesting

Reforestation need not be only used for recovery of accidentally destroyed forests. In some countries, such as Finland, the forests are *managed* by the wood products and pulp and paper industry. In such an arrangement, like other crops, trees are replanted wherever they are cut. In such circumstances, the industry can cut the trees in a way to allow easier reforestation. The wood products industry systematically replaces many of the trees it cuts, employing large numbers of summer workers for tree planting work. For example, in 2010, Weyerhaeuser reported planting 50 million seedlings.

In just 20 years, a teak plantation in Costa Rica can produce up to about 400 m of wood per hectare. As the natural teak forests of Asia become more scarce or difficult to obtain, the prices commanded by plantation-grown teak grow higher every year. Other species such as mahogany grow slower than teak in Tropical America but are also extremely valuable. Faster growers include pine, eucalyptus, and *Gmelina*.

Reforestation, if several native species are used, can provide other benefits in addition to financial returns, including restoration of the soil, rejuvenation of local flora and fauna, and the capturing and sequestering of 38 tons of carbon dioxide per hectare per year. The reestablishment of forests is not just simple tree planting. Forests are made up of a diversity of species and they build dead organic matter into soils over time. A major tree-planting programme in a place like this would enhance the local climate and reduce the demands of burning large amounts of fossil fuels for cooling in the summer.

For Climate Change Mitigation

Forests are an important part of the global carbon cycle because trees and plants absorb carbon dioxide through photosynthesis. By removing this greenhouse gas from the air, forests function as terrestrial carbon sinks, meaning they store large amounts of carbon. At any time, forests account for as much as double the amount of carbon in the atmosphere. Even as more anthropogenic carbon is produced, forests remove around three billion tons of anthropogenic carbon every year. This amounts to about 30% of all carbon dioxide emissions from fossil fuels. Therefore, an increase in the overall forest cover around the world would tend to mitigate global warming.

There are four major strategies available to mitigate carbon emissions through forestry activities: increase the amount of forested land through a reforestation process; increase the carbon density of existing forests at a stand and landscape scale; expand the use of forest products that will sustainably replace fossil-fuel emissions; and reduce carbon emissions that are caused from deforestation and degradation.

Achieving the first strategy would require enormous and wide-ranging efforts. However, there are many organizations around the world that encourage tree-planting as a way to offset carbon emissions for the express purpose of fighting climate change. For example, in China, the Jane Goodall Institute, through their Shanghai Roots & Shoots division, launched the Million Tree Project in Kulun Qi, Inner Mongolia to plant one million trees to stop desertification and help curb climate change. China has used 24 billion metres squared of new forest plantation and natural forest regrowth to offset 21% of Chinese fossil fuel emissions in 2000. In Java, Indonesia each newlywed couple is to give whoever is sermonizing their wedding 5 seedlings to combat global warming. Each couple that wishes to have a divorce has to give 25 seedlings to whoever divorces them.

The second stategy has to do with selecting species for tree-planting. In theory, planting any kind of tree to produce more forest cover would absorb more carbon dioxide from the atmosphere. On the other hand, a genetically modified tree specimen might grow much faster than any other regular tree. Some of these trees are already being developed in the lumber and biofuel industries. These fast-growing trees would not only be planted for those industries but they can also be planted to help absorb carbon dioxide faster than slow-growing trees.

Extensive forest resources placed anywhere in the world will not always have the same impact. For example, large reforestation programmes in boreal or subarctic regions have a limited impact on climate mitigation. This is because it substitutes a bright snow-dominated region that reflects the sunlight with dark forest canopies. A study from the National Centre for Atmospheric Research in Boulder, Colourado, USA, found that trees in temperate latitudes have a net warming effect on the atmosphere. The heat that dark leaves release without absorbing outweighs the carbon they sequester. On the other hand, a positive example would be reforestation projects in tropical regions, which would lead to a positive biophysical change such as the formation of clouds. These clouds would then reflect the sunlight, creating a positive impact on climate mitigation.

There is an advantage to planting trees in tropical climates with wet seasons. In such a setting, trees have a quicker growth rate because they can grow year-round. Trees in tropical climates have, on average, larger, brighter, and more abundant leaves than non-tropical climates. A study of the girth of 70,000 trees across Africa has shown that tropical forests are soaking up more carbon dioxide pollution than previously realized. The research suggests almost one fifth of fossil fuel emissions are absorbed by forests across Africa, Amazonia and Asia. Simon Lewis, a climate expert at the University of Leeds, who led the study, said: "Tropical forest trees are absorbing about 18% of the carbon dioxide added to the atmosphere each year from burning fossil fuels, substantially buffering the rate of change."

It is also important to deal with the rate of deforestation. At this point, there are 13 billion metres squared of tropical regions that are deforested every year. There is potential for these regions to reduce rates of deforestation by 50% by 2050, which would be a huge contribution to stabilize the global climate.

Incentives

Some incentives for reforestation can be as simple as a financial compensation. Streck and Scholz (2006) explain how a group of scientists from various institutions have developed a compensated reduction of deforestation approach which would reward developing countries that disrupt any further act of deforestation. Countries that participate and take the option to reduce their emissions from deforestation during a committed period of time would receive financial compensation for the carbon dioxide emissions that they avoided. To raise the payments, the host country would issue government bonds or negotiate some kind of loan with a financial institution that would want to take part in the compensation promised to the other country. The funds received by the country could be invested to help find alternatives to the extensive cutdown of forests. This whole process of cutting emissions would be voluntary, but once the country has agreed to lower their emissions they would be obligated to reduce their emissions. However, if a country was not able to meet their obligation, their target would get added to their next commitment period. The authors of these proposals see this as a solely government-to-government agreement; private entities would not participate in the compensation trades.

Examples

It is the stated goal of the US Forest Service to manage forest resources sustainably. This includes reforestation after timber harvest, among other programmes.

In Germany, reforestation is required as part of the federal forest law. 31% of Germany is forested, according to the second forest inventory of 2001–2003. The size of the forest area in Germany increased between the first and the second forest inventory due to forestation of degenerated bogs and agricultural areas. In China, extensive replanting programmes have existed since the 1970s. Programmes have had overall success. The forest cover has increased from 12% of China's land area to 16%. However, specific programmes have had limited success. The "Green Wall of China", an attempt to limit the expansion of the Gobi Desert is planned to be 2,800 miles (4,500 km) long and to be completed in 2050. In Canada, overall forest cover is increasing over the last decades.

In Borneo Dr Willie Smits, bought up nearly 2000 ha of deforested degraded land in East Kalimantan that had suffered from mechanical logging, drought and severe fires and was covered in alang-alang grass. In a project called Samboja Lestari an area was reforested.

The Groasis Waterboxx was designed specifically to establish trees in areas undergoing desertification. It collects dew and infrequent rain, and slowly releases it to the plants roots, promoting deeper root growth.

In many parts of the world, especially in East Asian countries, reforestation and afforestation are increasing the area of forested lands. The amount of woodland has increased in 22 of the world's 50 most forested nations. Asia as a whole gained 1 million hectares of forest between 2000 and 2005. Tropical forest in El Salvador expanded more than 20% between 1992 and 2001. Based on these trends, one study projects that global forest will increase by 10%—an area the size of India—by 2050.

In the People's Republic of China, where large scale destruction of forests has occurred, the government has in the past required that every able-bodied citizen between the ages of 11 and 60 plant three to five trees per year or do the equivalent amount of work in other forest services. The government claims that at least 1 billion trees have been planted in China every year since 1982. This is no longer required today, but March 12 of every year in China is the Planting Holiday. Also, it has introduced the Green Wall of China project, which aims to halt the expansion of the Gobi desert through the planting of trees. However, due to the large percentage of trees dying off after planting (up to 75%), the project is not very successful. There has been a 47-million-hectare increase in forest area in China since the 1970s. The total number of trees amounted to be about 35 billion and 4.55% of China's land mass increased in forest coverage. The forest coverage was 12% two decades ago and now is 16.55%.

An ambitious proposal for China is the Aerially Delivered Reforestation and Erosion Control System and the proposed Sahara Forest Project coupled with the Seawater Greenhouse.

In Western countries, increasing consumer demand for wood products that have been produced and harvested in a sustainable manner is causing forest landowners and forest industries to become increasingly accountable for their forest management and timber harvesting practices.

The Arbor Day Foundation's Rain Forest Rescue programme is a charity that helps to prevent deforestation. The charity uses donated money to buy up and preserve rainforest land before the lumber companies can buy it. The Arbor Day Foundation then protects the land from deforestation. This also locks in the way of life of the primitive tribes living on the forest land. Organizations such as Community Forestry International, Cool Earth, The Nature Conservancy, World Wide Fund for Nature, Conservation International, African Conservation Foundation and Greenpeace also focus on preserving forest habitats. Greenpeace in particular has also mapped out the forests that are still intact and published this information on the internet. World Resources Institute in turn has made a simpler thematic map showing the amount of forests present just before the age of man (8000 years ago) and the current (reduced) levels of forest. These maps mark the amount of afforestation required to repair the damage caused by people.

Forest Plantations

To meet the world's demand for wood, it has been suggested by forestry writers Botkins and Sedjo that high-yielding forest plantations are suitable. It has been calculated that plantations yielding 10 cubic metres per hectare annually could supply all the timber required for international trade on 5% of the world's existing forestland. By contrast, natural forests produce about 1–2 cubic metres per hectare; therefore, 5–10 times more forestland would be required to meet demand. Forester Chad Oliver has suggested a forest mosaic with high-yield forest lands interpersed with conservation land.

In the country of Senegal, on the western coast of Africa, a movement headed by youths has helped to plant over 6 million mangrove trees. The trees will protect local villages from storm damages and will provide a habitat for local wildlife. The project started in 2008, and already the Senegalese government has been asked to establish rules and regulations that would protect the new mangrove forests.

Military Context

While the preponderance of deforestation is due to demands for agricultural and urban use for the human population, there are some examples of military causes. One example of deliberate deforestation is that which took place in the U.S. zone of occupation in Germany after World War II. Before the onset of the Cold War, defeated Germany was still considered a potential future threat rather than potential future ally. To address this threat, attempts were made to lower German industrial potential, of which forests were deemed an element. Sources in the U.S. government admitted that the purpose of this was that the "ultimate destruction of the war potential of German forests." As a consequence of the practice of clear-felling, deforestation resulted which could "be replaced only by long forestry development over perhaps a century."

Deforestation can also be one consequence of war. For example, in the 1945 Battle of Okinawa, bombardment and other combat operations reduced the lush tropical landscape into "a vast field of mud, lead, decay and maggots". Deforestation can also be an intentional tactic of military forces. Defoliants (like Agent Orange or others) was used by the British in the Malayan Emergency, and by the United States in the Korean War and Vietnam War.

Afforestation

Afforestation is the establishment of a forest or stand of trees in an area where there was no forest. Reforestation is the reestablishment of forest cover, either naturally (by natural seeding, coppice, or root suckers) or artificially (by direct seeding or planting). Many governments and non-governmental organizations directly engage in programmes of *afforestation* to create forests, increase carbon capture and sequestration, and help to anthropogenically improve biodiversity. (In the UK, afforestation may mean converting the legal status of some land to "royal forest".) Special tools, e.g. tree planting bar, are used to make planting of trees easier and faster.

Biological process

Gap dynamics refers to the pattern of plant growth that occurs following the creation of a forest gap, a local area of natural disturbance that results in an opening in the canopy of a forest. Gap dynamics are a typical characteristic of both temperate and tropical forests and have a wide variety of causes and effects on forest life.

Gap Dynamics

Gap dynamics refers to the pattern of plant growth that occurs following the creation of a forest gap, a local area of natural disturbance that results in an opening in the canopy of a forest. Gap dynamics are a typical characteristic of both temperate and tropical forests and have a wide variety of causes and effects on forest life.

Gaps are the result of natural disturbances in forests, ranging from a large branch breaking off and dropping from a tree, to a tree dying then falling over, bringing its roots to the surface of the ground, to landslides bringing down large groups of trees. Because of the range of causes, gaps therefore have a wide range of sizes, including small and large gaps.

Regardless of size, gaps allow an increase in light as well as changes in moisture and wind, leading to differences in microclimate conditions compared to those from below the closed canopy, which are generally cooler and more shaded.

For gap dynamics to occur in naturally disturbed areas, either primary or secondary succession must occur. Ecological secondary succession is much more common and pertains to the process of vegetation replacement after a natural disturbance. Secondary succession results in second-growth or secondary forest, which currently covers more of the tropics than old-growth forest.

Since gaps let in more light and create diverse microclimates, they provide the ideal location and conditions for rapid plant reproduction and growth. In fact, most plant species in the tropics are dependent, at least in part, on gaps to complete their life cycles.

Disturbances

Gap dynamics are the result of disturbances within an ecosystem. There are both large scale and small scale disturbances, and both are influenced by duration and frequency. These all affect the resulting impact and regeneration patterns of the ecosystem.

The most common type of disturbance within a tropical ecosystem is fire. Since most nutrients in a tropical ecosystem are contained in the biomass of plants, fire is an important component of recycling these nutrients and therefore regenerating an ecosystem. An example of a small scale disturbance is a tree falling. This can cause soil movement, which redistributes any nutrients or organisms that were attached to the tree. The tree falling also opens up the canopy for light entrance, which can support the growth of other trees and plants.

After a disturbance, there are several ways in which regeneration can occur. One way, termed the advance regeneration pathway, is when the primary understory already contains seedlings and saplings. This method is most common in the Neotropics when faced when small scale disturbances. The next pathway is from tree remains, or any growth from bases or roots, and is common in small disturbance gaps. The third route is referred to as the soil seed bank, and is the result of germination of seeds already found in the soil. The final regeneration pathway is the arrival of new seeds via animal dispersal or wind movement. The most critical components of the regeneration are seed distribution, germination, and survival.

Primary Succession

Succession is the slow rebuilding of forest gaps from natural or human disturbances. When major geological changes such as volcano eruptions or landslides occur, the current vegetation and soil may erode away leaving only rock. Primary succession occurs when pioneer species such as lichens colonize rock. As the lichens and mosses decompose, a soil substrate forms called peat. The peat, over time, will create a terrestrial ecosystem. From there on herbaceous, non-woody plants will develop and trees will follow. Major holes or gaps in the forest ecosystem will take hundreds of years to regenerate from a rock base.

Secondary Succession

Secondary succession occurs where a disturbance has taken place but soil remains and is able to support plant growth. It does not take nearly as long for plant regeneration to occur because of the soil substrate already present. Secondary succession is much more common than primary succession in the tropics.

Ecological secondary succession occurs in four distinct phases: First, rapid colonization of cleared land by species such as herbs, shrubs, and climbers as well as seedlings from pioneer tree species occurs and this can last up to three years. After that, short lived but fast growing shade intolerant species form a canopy over 10 to 30 years. Non-pioneer heliophilic (sun-loving) tree species then add to the biomass and species richness as well as shade tolerant species and this can last 75 to 150 years. Finally, shade-tolerant species regain full canopy stature indefinitely until another major disturbance occurs.

Secondary succession in the tropics begins with pioneer species, which are rapidly growing and include vines and shrubs. Once these species are established, large heliophilic species will develop such as heliconias. Cecropias are also a major pioneering tree in the tropics and they are adapted to grow well where forest gaps are giving way to sunlight. Shade-tolerant species that have remained low in the forest develop and become much taller. These successional phases do not have definite order or structure and because of the very high biodiversity in the tropics, there is a lot of competition for resources such as soil nutrients and sunlight.

Examples of tree dynamics

Due to the fact that horizontal and vertical heterogeneity of a forest is significantly increased by gaps, gaps become an obvious consideration in explaining high biodiversity. It has been proven that gaps create suitable conditions for rapid growth and reproduction. For example, non-shade tolerant plant species and many shade-tolerant plant species respond to gaps with an increase in growth, and at least a few species are dependent on gaps to succeed in their respected environments (Brokaw 1985; Hubbell and Foster 1986b; Murray 1988; Clark and Clark 1992). Gaps create diverse microclimates, affecting light, moisture, and wind conditions (Brokaw 1985). For example, exposure to edge effects increases a microclimate's light and wind intensity and decreases its moisture. A study conducted on Barro Colorado Island in Panama showed that gaps had greater seedling establishment and higher sapling densities than control areas.

Species richness was higher in gaps than in control areas, and there was more diversity in species composition among gaps. However, this study also found that there was a low recruitment rate per gap, which explains why gaps differed in species composition. With 2% to 3% for pioneer species and 3% to 6% for shade-tolerant and intermediate species. Suggesting that most species could not take advantage of gaps because they couldn't get to them through seed dispersal. With that said, the Janzen-Connell effect plays a major role in the tree species' relationship with gaps. The Janzen-Connell density dependent mortality model states that most trees die as seed or seedlings. In addition, host-specific predators or pathogens are predicted to be greatest where density is greatest, which is underneath parent tree. This corroborates with the major causes of gaps, which are the falling of trees due to mortality caused by termites or epiphyte growth. The Janzen-Connell model also states that balance between dispersal distance and mortality should cause highest recruitment to be at a certain distance away from the parent. Therefore if these gaps are being created by the parents, the seedlings recruit away from the gap, resulting in increasing survival rates as the distance from the parent increases. This explains the low recruitment rate per gap found in the experiment conducted in Barro Colorado Island.

In corroboration, a study conducted in La Selva in Costa Rica calculated the crown illumination index for nine tree species ranging from gap specialists to emergent canopy species. Crown illumination values ranged from 1, which indicated low light, and 6, which indicated that the tree crown was completely exposed . After using a mathematical model to calculate the changes in tree diametre and changes in crown illumination with age. This model helped estimate life expectancy, time of passage to various sizes, and age patterns of mortality. The results showed what most gap dynamics studies show, pioneer species thrived in high light environments and non-pioneer species showed high mortality when young but the rate of mortality decreased as they aged. However, once trees were very large survivorship then decreased.

In areas of Degraded Soil

In some places, forests need help to reestablish themselves because of environmental factors. For example, in arid zones, once forest cover is destroyed, the land may dry and become inhospitable to new tree growth. Other factors include overgrazing by livestock, especially animals such as goats, cows, and over-harvesting of forest resources. Together these may lead to desertification and the loss of topsoil; without soil, forests cannot grow until the long process of soil creation has been completed - if erosion allows this. In some tropical areas, forest cover removal may result in a duricrust or duripan that effectively seal off the soil to water penetration and root growth. In many areas, reforestation is impossible because people are using the land. In other areas, mechanical breaking up of duripans or duricrusts is necessary, careful and continued watering may be essential, and special protection, such as fencing, may be needed.

In areas of extremely poor soil, the Groasis Waterboxx has been effective in growing young trees. The Groasis Waterboxx was designed specifically to establish trees in areas undergoing desertification. It collects dew and infrequent rain, and slowly releases it to the plants roots, promoting deeper root growth.

Countries and Regions

Brazil: Because of the extensive Amazon deforestation during the last decades and ongoing, the small efforts of afforestation are insignificant on a national scale of the Amazon Rainforest.

China: China has deforested most of its historically wooded areas. China reached the point where timber yields declined far below historic levels, due to over-harvesting of trees beyond sustainable yield. Although it has set official goals for reforestation, these goals were set for an 80 year time horizon and are not significantly met by 2008. China is trying to correct these problems by projects as the Green Wall of China, which aims to replant a great deal of forests and halt the expansion of the Gobi desert. A law promulgated in 1981 requires that every citizen over the age of 11 plant at least one tree per year. As a result, China currently has the highest afforestation rate of any country or region in the world, with 47,000 square kilometres of afforestation in 2008. However, the forest area per capita is still far lower than the international average.

North Africa: In North Africa, the Sahara Forest Project coupled with the Seawater greenhouse has been proposed. Some projects have also been launched in countries as Senegal to revert desertification. As of 2010, African leaders are discussing the combining of national resources to increase effectiveness. In addition, other projects as the Keita Project in Niger have been launched in the past, and have been able to locally revert damage done by desertification.

Europe: Europe has deforested the majority of its historical forests. The European Union (EU) has paid farmers for afforestation since 1990, offering grants to turn farmland back into forest and payments for the management of forest. Between 1993 and 1997, EU afforestation policies made possible the re-forestation of over 5,000 square kilometres of land. A second programme, running between 2000 and 2006, afforested more than 1000 square kilometres of land (precise statistics not yet available). A third such programme began in 2007.

In Poland, the National Programme of Afforestation was introduced by the government after World War II, when area of forests shrank to 20% of country's territory. Consequently, forested areas of Poland grew year by year, and on December 31, 2006, forests covered 29% of the country. It is planned that by 2050, forests will cover 33% of Poland.

According to FAO statistics, Spain had the third fastest afforestation rate in Europe in the 1990-2005 period, after Iceland and Ireland. In those years, a total of 44,360 square kilometres were afforested, and the total forest cover rose from 13,5 to 17,9 million hectares. In 1990, forests covered 26,6% of the Spanish territory. As of 2007, that figure had risen to 36,6%. Spain today has the fifth largest forest area in the European Union.

In January 2013 the UK government set a target of 12% woodland cover in England by 2060, up from the then 10%. Government-backed initiatives such as the Woodland Carbon Code are intended to support this objective by encouraging corporations and landowners to create new woodland to offset their carbon omissions.

Iran: Iran is considered a low forest cover region of the world with present cover approximating seven percent of the land area. This is a value reduced by an estimated six million hectares of virgin forest, which includes oak, almond and pistachio. Due to soil substrates, it is difficult to achieve afforestation on a large scale compared to other temperate areas endowed with more fertile and less rocky and arid soil conditions. Consequently, most of the afforestation is conducted with non-native species, leading to habitat destruction for native flora and fauna, and resulting in an accelerated loss of biodiversity.

India: India, after 1950 till 2006 has witnessed a minor increase in the percentage of the land area under forest cover. In 1950 around 40.48 million hectare area was under forest cover. in 1980 it increased to 67.47 million hectare and in 2006 it was found to be 69 million hectare. Out of the total land available around 23% of land is under forest cover. The forests in India have been grouped into 5 major categories and 16 types according to biophysical criteria. The distribution of these groups indicates 38.20% subtropical dry deciduous, 30.30% tropical moist deciduous, 6.7% subtropical thorn and 5.8% tropical wet evergreen forests. Other categories include subtropical pine (5%), tropical semi-evergreen forests (2.5%) and other smaller categories. Temperate and alpine areas cover about 10% of the forest areas. It is taken care that only local species are planted in an area. Trees bearing fruits are of higher choice in any geography.

Afforestation in Japan

The Japanese temperate rainforest is well sustained and maintains a high biodiversity. One method that has been utilized in maintaining the health of forests in Japan has been afforestation. The Japanese government and private businesses have set up multiple projects to plant native tree species in open areas scattered throughout the country. This practice has resulted in shifts in forest structure and a healthy temperate rainforest that maintains a high biodiversity.

Purpose of Afforestation

The primary goal of afforestation projects in Japan is to develop the forest structure of the nation and to maintain the biodiversity found in the Japanese wilderness. The Japanese temperate rainforest is scattered throughout the Japanese archipelago and is home to many endemic species that are not naturally found anywhere else. As development of the country's caused a decline in forest cover, a reduction in biodiversity was seen in those areas. In an effort to counteract the observed decline in biodiversity, Japan began many afforestation projects. New tree stands were planted all over the archipelago and native species that inhabited the existing wild forests began to occupy the newly forested areas.

Afforestation projects in Japan first started after the rebuilding that followed World War II. In efforts to restore the country's infrastructure after the war, large areas of forest were clear-cut for timber and to create pastures to attract immigrant farmers. A new management plan for the forests of Japan was instated after many pastures were abandoned and there was a recognized massive decline of old growth and secondary forests. Forest plantations were created to increase the health of Japanese forests and to sustain the nation's timber industry. Afforestation was combined with changes in logging practices that called for reduced clear-cutting and low impact logging over a larger area.

Many private businesses in the country take part in other afforestation projects as a means of reducing the carbon emissions of the company. Carbon sequestration is a major incentive for businesses to plant seedlings and saplings that will store atmospheric carbon in their biomass as they grow. Companies like Japex and Toyota have planted and maintained tree stands beside their plants located across Japan; planting several thousand native trees in an effort to offset carbon emissions. Businesses also monitor the health of newly planted tree stands by tracking growth and surveying near-by wild forests as a comparison of how well the new stands are doing.

Historically, the forests of Japan were not extensively cut as a means of reducing the frequency of landslides and other natural disasters occurring. The root structure of the forests held the soil in place and stabilized the ground in an environment that experiences heavy rainfall and earthquake activity. In present afforestation projects, trees are used to anchor the soil and reduce the amount of soil erosion undergone by that area. A healthy tree stand also restricts the amount of soil disturbance due to rainfall by intercepting heavy precipitation in the canopy and by slowing the rate of run-off with the litter covering the soil. The role that the Japanese temperate rainforest plays in the prevention of soil erosion was recognized early on in the history of the nation and continues to be a positive reason to reforest open areas as new tree stands reduce the amount of nutrients lost in an area and allow for a more productive ecosystem.

Processes Involved

There has been varying degrees of human involvement in afforested regions. For most of the areas that are reforested, there is a clearing of land that is planted with native saplings. Growth is generally monitored, but there is not much human involvement beyond checking the health of a stand.

In other regions, there is more land preparation involved prior to planting anything. Most open areas get quickly colonized by dwarf bamboo that out-competes any saplings trying to grow in the area. To give any new saplings a chance to grow in a new area people practice soil scarification. The land is plowed and soil is rotated. This process kills off any tall grasses of bamboo in the area, effectively reducing competition for new saplings. This process does subject the soil to compaction, making it harder for roots to establish; however, once the trees are established, they can regenerate and facilitate growth of other tree species. To facilitate further growth and development in plantations, there is selective logging in order to create microhabitats for new growth using pit-and-mound topography. Trees are tipped over to create mounds that turn up more nutrients in the soil and allow for more sunlight in some patches of the understory. A healthy understory of young trees serves as a source of regeneration when an older, larger tree dies.

Tree Species

Japan afforestation projects plant only tree species native to Japan. Originally, species were chosen for their rapid growth rates and tolerance for multiple environmental conditions, which lead to a large shift from old growth and secondary forests to conifer plantations. Current afforestation projects plant a more diverse number of species. The dominant tree species planted are Japanese cedar (*Cryptomeria japonica*) and Japanese cypress (*Chamaecyparis obtusa*). Both conifer species are native to Japan and prefer to grow in deep, well-drained soils in warm, wet climates. Both species were able to successfully occupy the open areas made available by rapid deforestation. Other than being able to grow quickly, the trees were chosen because of their ornamental use in urban areas and because their wood is highly valuable timber.

Other native tree species are planted in an effort to regenerate mixed forests of conifers and broadleaf deciduous trees. Other species include: Mongolian oak (*Quercus mongolica*), lobed elm (*Ulmus laciniata*), Japanese white birch (*Betula platyphylla*), Glehn's spruce (*Picea glehnii*), and sakhalin fir (*Abies sachalinensis*).

Changes in Soil

One study examining the soil of various Japanese conifer plantations found that the plantations had a higher concentration of soil organic carbon than wild, natural forests in Japan. Soil carbon is associated with the amount of nutrients that a soil sample can hold. When areas undergo deforestation a lot of the soil organic carbon is exposed and vulnerable to degradation. This results in a loss of nutrients in deforested areas. It would be expected that plantations would have a much smaller amount of organic carbon in the soil because the soil was exposed and experienced a nutrient loss; however, it was found that conifer plantations in Japan actually had more soil carbon than wild forests in the area. It was determined that the heavy litter from newer conifer plantations resulted in an abundance of soil carbon. It was also noted that there was less soil carbon found in plantations that had undergone soil scarification prior to being planted. This was determined to be because soil scarification brought up and exposed more soil organic carbon and more was lost to degradation; however there was still a high concentration of soil carbon in those plantations as well.

Concerns

Afforestation is generally seen as a positive thing, and the re-establishment of forests in Japan has been positive overall. Large-scale afforestation is still a new concept, and there are some concerns associated with it. Afforested regions have impacts on local watersheds as well as species interactions.

One study examined the reduced water quality of watersheds downstream of afforested conifer plantations. The concern was that there was no improvement seen with public water despite improved water treatment in those areas. It was found that heavy rainfall upstream in afforested conifer plantations caused an increase in nitrogen and phosphorus running off into the streams. The undeveloped understory in those plantations means that a lot of the soil is exposed. The litter does slow down the rate of the run-off, but a large amount of nutrients still get picked up and put into the watershed. This resulted in an increase in particulate and dissolved nitrogen and phosphorus.

A large concern in Japan is root rot experienced in conifer species as a result of *Armillaria*, a group of fungus species found in Asia. Sites all over Japan were examined and it was found that every conifer species native to Japan was vulnerable to at least one species of *Armillaria*. This is a major concern because every conifer species is vulnerable, but it is especially concerning because the majority of the Japanese temperate rainforest is made up of conifer plantations consisting of monocultures of just a few species.

Another concern is due to Japan's practice of importing timber from other places. Although this has proven to be beneficial for Japanese forests, it has resulted in massive deforestation in places that do not have well-developed regulation on logging. Japan is one of the largest importers of timber, which has come at a cost of temperate and tropical forest decline in China and South America. As a result of this negativity, Japan has organized afforestation programmes in other countries.

Buffer Strip

A buffer strip is an area of land maintained in permanent vegetation that helps to control air, soil, and water quality, along with other environmental problems, dealing primarily on land that is used in agriculture. Buffer strips trap sediment, and enhance filtration of nutrients and pesticides by slowing down runoff that could enter the local surface waters. The root systems of the planted vegetation in these buffers hold soil particles together which alleviate the soil of wind erosion and stabilize stream banks providing protection against substantial erosion and landslides. Farmers can also use buffer strips to square up existing crop fields to provide safety for equipment while also farming more efficiently.

Buffer strips can have several different configurations of vegetation found on them varying from simply grass to combinations of grass, trees, and shrubs. Areas with diverse vegetation provide more protection from nutrient and pesticide flow and at the same time provide better biodiversity amongst plants and animals.

Many country, state, and local governments provide financial incentives for conservation programmes such as buffer strips because they help stabilize the environment even when the land is being used.

Buffer strips not only stabilize the land but can also provide a visual demonstration that land is under stewardship.

Types of Buffer Strips

Buffers within Fields

Grassed waterway: A grassed waterway reduces soil erosion and captures most nutrients and pesticides that would normally wash out of crop fields and into major waters. These waterways help to carry surface water at a non-erosive velocity to an area where it will have a stable outlet. Outlets must be adequate enough to allow water to drain without ponding or flooding the area being protected, while also preventing erosion of the water into the outlet which can be accomplished through the use of riprap. A limitation is during large runoff events, when soil is saturated, grassed waterways will have a very concentrated flow of water making them not as effective during high rainfalls. Grassed waterways require very little maintenance once they are introduced with major upkeep being mowing of the grass and reseeding. Farm machinery and cattle can cross these waterways but it may be hazardous during wet periods. One of the major disadvantages of waterways are actually getting them established. A late summer or early fall seeding when rainfall is minimal is recommended to allow the seed to have the best chance at establishing a root system.

Contour Buffer Strips

This type of buffer strip is simply a strip of perennial vegetation that is alternated with wider cultivated strips of cropland. This type of buffer strip is most effective when runoff water enters uniformly as sheetflow. They are very adapted to trapping pesticides and reducing rill erosion. These buffers need to be at least 15 feet (4.6 m) wide and make up for 20 to 30 percent of slope of an area. A lot of the time contour buffers can be used as a very inexpensive substitute for terraces. Most of the time a grass is selected that can be harvested during mid-summer. These buffers are not permanent and can be moved up and down hillsides from season to season in order to re-establish vegetation.

Edge of Field Buffers

Vegetative barriers: This buffer strip is much narrower than most of the other types of buffers. They are narrow, permanent strips of hardy, native, perennial grasses or shrubs planted in parallel rows to crops. They are very affective in reducing wind and water erosion which results in sediment trapping and water infiltration. They function in similar fashion to a contour buffer strip, just much narrower.

Field Borders

This type of buffer strip is simply a band or strip of perennial vegetation that is found on the edge of a cropland field. Field borders help with runoff only when it flows over the strip. They're very effective in benefiting spraying operations because they allow for extra room between adjacent fields. They also provide room for farming equipment to turn around. Field borders are effective in reducing wind and water erosion and provide great wildlife habitat.

Filter Strips

These strips are areas of grass or other permanent vegetation that protect riparian areas from sediment runoff, pesticides, pathogens, organics and nutrients. These strips are very important in protecting stream banks and water quality. Filter strips work best when other conservation practices are used in order to drain water in their direction. Filter strips were originally used mostly in agriculture, but now are a common practice in urban areas where water quality has become an increasingly important issue.

Riparian Forest Buffer

Figure: *A riparian buffer of vegetation lining a farm creek in Story County, Iowa.*

This type of buffer has a very diverse community of trees, shrubs and native perennial grasses. They are great for providing habitat for wildlife on land and in the water. Taller trees next to the streams help to lower water temperatures with shade which improves aquatic communities. The shrubs and grasses help to slow flooding and the larger trees can sometimes intercept nitrates before they reach the water with their deep roots.

Wind Buffers

The main purpose of a windbreak or a shelterbelt is to protect areas from wind causing erosion on the bare soil of croplands. Windbreaks can also serve as an area that separates fields and protects them from spray drift of pesticides.

Wildlife Habitat

Buffer strips are very important in helping to provide habitat for many species of wildlife in the open farm lands by causing an edge effect. With much of the land open on farms having an edge allows a safe-haven for animals to move between different ecosystems. Buffers are also helpful in conserving biodiversity especially to that of rare or endangered species through the incorporation of native grasses into their seeding by the landowner.

Native grasses are especially important to pheasants, quail, chukar and songbirds because they provide the foods they need as well as the shelter from predators. Since most buffer strip areas have limited disturbance from farming it allows for a shelter to hide year round for many of the species including insects, birds, and mammals. When buffers run into and follow riparian areas along stream beds it is important to have larger vegetation like trees and shrubs that shade the water from the open sun. The water is then able to be cooler allowing for greater fish production and other aquatic plants and other biota to thrive in a less disturbed environment.

This area of vegetation following a body of water is noted as a riparian zone. These buffer areas often incorporate large trees that protect stream banks from excessive erosion and shade aquatic areas. The shade provided by the larger trees reduces water temperatures and light intensity from ultraviolet light.

Debris including leaves and branches that fell from trees, often contain aquatic invertebrates important to the structure of the water following the "River Continuum" concept. Since the riparian zones contain a larger variety of plants the overall diversity is much greater as well. With more photosynthesis and higher amounts of available water plant primary production can increase in turn creating more potential food for the wildlife

Planting Vegetation

Grassed waterways: The USDA shows that the use of less than 5% slopes for a chosen waterway will greatly reduce velocity of draining water within the land. The NRCS offers the use of an online Soil survey to view the area of land to be planted and examine slope and soil drainage. As viewed by the NRCS, soil drainage is the number one priority for location of certain grasses. Poorer drainage causes less infiltration of water into the soil as groundwater recharge causing ponding and flooding of surrounding crops. Higher drainage becomes more droughty which hinders the establishment of certain grasses good for waterways. Medium soil drainage becomes the most suitable for planting. Determining exact grasses and seed amounts to plant follows specifications of a local NRCS Field Office Technical Guide and the Contour Buffer practice standard while taking practice on land and regional environmental conditions into consideration.

Contour Strip Cropping

The NRCS has shown contour strips have the greatest effectiveness when slopes of the area are within 4-8%. For greater success in lowering the erosion, contour strips on the slopes need to follow the contours of the landscape. Row crops like corn, legumes, or soybeans should alternate strips of small grains or forages to successfully limit erosion and slowing or capturing the runoff of fertilizers and pesticides. Tests from the NRCS have shown that the smaller tighter rooted grasses and hays provide more surface cover to prevent rain and wind erosion while slowing runoff, increasing infiltration, and trapping sediment from the high erosive row crops. Proper planting of the contour crops is important for higher success of erosion control to protect highly fertile topsoil.

Strategies towards Desertification Control in the Thar Desert

Desertification is a type of land degradation in which a relatively dry land region becomes increasingly arid, typically losing its bodies of water as well as vegetation and wildlife. It is caused by a variety of factors, such as climate change and human activities. Desertification is a significant global ecological and environmental problem.

Considerable controversy exists over the proper definition of the term "desertification" for which Helmut Geist (2005) has identified more than 100 formal definitions.The most widely accepted of these is that of the Princeton University Dictionary which defines it as "the process of fertile land transforming into desert typically as a result of deforestation, drought or improper/inappropriate agriculture"

The earliest known discussion of the topic arose soon after the French colonization of West Africa, when the Comité d'Etudes commissioned a study on *dessèchement progressif* to explore the prehistoric expansion of the Sahara Desert.

History

The world's most noted deserts have been formed by natural processes interacting over long intervals of time. During most of these times, deserts have grown and shrunk independent of human activities. Paleodeserts are large sand seas now inactive because they are stabilized by vegetation, some extending beyond the present margins of core deserts, such as the Sahara, the largest hot desert.

Desertification has played a significant role in human history, contributing to the collapse of several large empires, such as Carthage, Greece, and the Roman Empire, as well as causing displacement of local populations.

Historical evidence shows that the serious and extensive land deterioration occurring several centuries ago in arid regions had three epicentres: the Mediterranean, the Mesopotamian Valley, and the loessial plateau of China, where population was dense.

Areas Affected

Drylands occupy approximately 40–41% of Earth's land area and are home to more than 2 billion people. It has been estimated that some 10–20% of drylands are already degraded, the total area affected by desertification being between 6 and 12 million square kilometres, that about 1–6% of the inhabitants of drylands live in desertified areas, and that a billion people are under threat from further desertification.

As of 1998, the then-current degree of southward expansion of the Sahara was not well known, due to a lack of recent, measurable expansion of the desert into the Sahel at the time.

Vegetation Patterning

As the desertification takes place, the landscape may progress through different stages and continuously transform in appearance. On gradually sloped terrain, desertification can create increasingly larger empty spaces over a large strip of land, a phenomenon known as "Brousse tigrée".

A mathematical model of this phenomenon proposed by C. Klausmeier attributes this patterning to dynamics in plant-water interaction. One outcome of this observation suggests an optimal planting strategy for agriculture in arid environments.

Causes

The immediate cause is the removal of most vegetation. This is driven by a number of factors, alone or in combination, such as drought, climatic shifts, tillage for agriculture, overgrazing and deforestation for fuel or construction materials. Vegetation plays a major role in determining the biological composition of the soil. Studies have shown that, in many environments, the rate of erosion and runoff decreases exponentially with increased vegetation cover. Unprotected, dry soil surfaces blow away with the wind or are washed away by flash floods, leaving infertile lower soil layers that bake in the sun and become an unproductive hardpan. Alternatively, recent research has suggested that the movement and migration of large herds of livestock and wildlife has an integral role in the preservation of vegetation and soil fertilization, and that the removal of livestock and wildlife (largely by human influence) has been the main driver of increasing desertification.

Poverty

At least 90% of the inhabitants of drylands live in developing nations, where they also suffer from poor economic and social conditions. This situation is exacerbated by land degradation because of the reduction in productivity, the precariousness of living conditions and the difficulty of access to resources and opportunities.

A downward spiral is created in many underdeveloped countries by overgrazing, land exhaustion and overdrafting of groundwater in many of the marginally productive world regions due to overpopulation pressures to exploit marginal drylands for farming. Decision-makers are understandably averse to invest in arid zones with low potential. This absence of investment contributes to the marginalisation of these zones. When unfavourable agro-climatic conditions are combined with an absence of infrastructure and access to markets, as well as poorly adapted production techniques and an underfed and undereducated population, most such zones are excluded from development.

Desertification often causes rural lands to become unable to support the same sized populations that previously lived there. This results in mass migrations out of rural areas and into urban areas, particularly in Africa. These migrations into the cities often cause large numbers of unemployed people, who end up living in slums.

Countermeasures and Prevention

Techniques exist for mitigating or reversing the effects of desertification, however there are numerous barriers to their implementation. One of these is that the costs of adopting sustainable agricultural practices sometimes exceed the benefits for individual farmers, even while they are socially and environmentally beneficial. Another issue is a lack of political will, and lack of funding to support land reclamation and anti-desertification programmes.

Desertification is recognized as a major threat to biodiversity. Some countries have developed Biodiversity Action Plans to counter its effects, particularly in relation to the protection of endangered flora and fauna. Reforestation gets at one of the root causes of desertification and isn't just a treatment of the symptoms. Environmental organizations work in places where deforestation and desertification are contributing to extreme poverty. There they focus primarily on educating the local population about the dangers of deforestation and sometimes employ them to grow seedlings, which they transfer to severely deforested areas during the rainy season.

Techniques focus on two aspects: provisioning of water, and fixation and hyper-fertilizing soil.

Fixating the soil is often done through the use of shelter belts, woodlots and windbreaks. Windbreaks are made from trees and bushes and are used to reduce soil erosion and evapotranspiration. They were widely encouraged by development agencies from the middle of the 1980s in the Sahel area of Africa.

Some soils (for example, clay), due to lack of water can become consolidated rather than porous (as in the case of sandy soils). Some techniques as zaï or tillage are then used to still allow the planting of crops. Another technique that is useful is contour trenching. This involves the digging of 150m long, 1m deep trenches in the soil. The trenches are made parallel to the height lines of the landscape, preventing the water of flowing within the trenches and causing erosion. Stone walls are placed around the trenches to prevent the trenches of closing up again. The method was invented by Peter Westerveld.

Enriching of the soil and restoration of its fertility is often done by plants. Of these, the Leguminous plants which extract nitrogen from the air and fixes it in the soil, and food crops/trees as grains, barley, beans and dates are the most important. Sand fences can also be used to control drifting of soil and sand erosion.

As there are many different types of deserts, there are also different types of desert reclamation methodologies. An example for this is the salt-flats in the Rub' al Khali desert in Saudi-Arabia. These salt-flats are one of the most promising desert areas for seawater agriculture and could be revitalized without the use of freshwater or much energy.

Farmer-managed natural regeneration (FMNR) is another technique that has produced successful results for desert reclamation. Since 1980, this method to reforest degraded landscape has been applied with some success in Niger. This simple and low-cost method has enabled farmers to regenerate some 30,000 square kilometres in Niger.

The process involves enabling native sprouting tree growth through selective pruning of shrub shoots. The residue from pruned trees can be used to provide mulching for fields thus increasing soil water retention and reducing evaporation. Additionally, properly spaced and pruned trees can increase crop yields. The Humbo Assisted Regeneration Project which uses FMNR techniques in Ethiopia has received money from The World Bank's BioCarbon Fund, which supports projects that sequester or conserve carbon in forests or agricultural ecosystems.

Managed Grazing

Restoring grasslands store CO_2 from the air into plant material. Grazing livestock, usually not left to wander, would eat the grass and would minimize any grass growth while grass left alone would eventually grow to cover its own growing buds, preventing them from photosynthesizing and killing the plant. A method proposed to restore grasslands uses fences with many small paddocks and moving herds from one paddock to another after a day a two in order to mimick natural grazers and allowing the grass to grow optimally.

It is estimated that increasing the carbon content of the soils in the world's 3.5 billion hectares of agricultural grassland would offset nearly 12 years of CO_2 emissions. Allan Savory, as part of holistic management, claims that while large herds are often blamed for desertification, prehistoric lands used to support large or larger herds and areas where herds were removed in the United States are still desertifying.

5

Fluorescence of Biologically Active Compounds in Plant Secretory Cells

Fluorescence is the emission of light by a substance that has absorbed light or other electromagnetic radiation. It also occurs when molecules are excited to higher electronic states by energetic electron bombardment, such as occurs, for example, in the natural aurora, high-altitude nuclear explosions, and rocket-born electron gun experiments. It is a form of luminescence. In most cases, the emitted light has a longer wavelength, and therefore lower energy, than the absorbed radiation. However, when the absorbed electromagnetic radiation is intense, it is possible for one electron to absorb two photons; this two-photon absorption can lead to emission of radiation having a shorter wavelength than the absorbed radiation. The emitted radiation may also be of the same wavelength as the absorbed radiation, termed “resonance fluorescence”.

The most striking examples of fluorescence occur when the absorbed radiation is in the ultraviolet region of the spectrum, and thus invisible to the human eye, and the emitted light is in the visible region. Fluorescence has many practical applications, including mineralogy, gemology, chemical sensors (fluorescence spectroscopy), fluorescent labelling, dyes, biological detectors, and, most commonly, fluorescent lamps.

Plant Secretory Tissue

The tissues that are concerned with the secretion of gums, resins, volatile oils, nectar latex, and other substances in plants are called secretory tissues These tissues are divided into two groups-

1. Laticiferous tissues
2. Glandular tissues

Cells or organizations of cells which produce a variety of secretions. The secreted substance may remain deposited within the secretory cell itself or may be excreted, that is, released from the cell. Substances may be excreted to the surface of the plant or into intercellular cavities or canals. Some of the many substances contained in the secretions are not further utilized by the plant (resins, rubber, tannins, and various crystals), while others take part in the functions of the plant (enzymes and hormones). Secretory structures range from single cells scattered among other kinds of cells to complex structures involving many cells; the latter are often called glands.

Epidermal hairs of many plants are secretory or glandular. Such hairs commonly have a head composed of one or more secretory cells borne on a stalk. The hair of a stinging needle is bulbous below and extends into a long, fine process above. If one touches the hair, its tip breaks off, the sharp edge penetrates the skin, and the poisonous secretion is released. Glands secreting a sugary liquid—the nectar—in flowers pollinated by insects are called nectaries. Nectaries may occur on the floral stalk or on any floral organ: sepal, petal, stamen, or ovary.

The hydathode structures discharge water—a phenomenon called guttation through openings in margins or tips of leaves. The water flows through the xylem to its endings in the leaf and then through the Intercellular spaces of the hydathode tissue towards the openings in the epidermis. Strictly speaking, such hydathodes are not glands because they are passive with regard to the flow of water.

Some carnivorous plants have glands that produce secretions capable of digesting insects and small animals. These glands occur on leaf parts modified as insect-trapping structures. In the sundews (Drosera) the traps bear stalked glands, called tentacles. When an insect lights on the leaf, the tentacles bend down and cover the victim with a mucilaginous secretion, the enzymes of which digest the insect.

Resin ducts are canals lined with secretory cells that release resins into the canal. Resin ducts are common in gymnosperms and occur in various tissues of roots, stems, leaves, and reproductive structures.

Gum ducts are similar to resin ducts and may contain resins, oils, and gums. Usually, the term gum duct is used with reference to the dicotyledons, although gum ducts also may occur in the gymnosperms. Oil ducts are intercellular canals whose secretory cells produce oils or similar substances. Such ducts may be seen, for example, in various parts of the plant of the carrot family (Umbelliferae). Laticifers are cells or systems of cells containing latex, a milky or clear, coloured or colourless liquid. Latex occurs under pressure and exudes from the plant when the latter is cut.

Laticiferous Tissues

These consist of thin walled, greatly elongated and much branched ducts containing a milky or yellowish coloured juice known as latex. They contain numerous nuclei which lie embedded in the thin lining layer of protoplasm. They irregularly distributed in the mass of parenchymatous cells. Laticiferous ducts, in which latex are found are again two types-

1. Latex cell or non-articulate latex ducts
2. Latex vessels or articulate latex

Latex Cells

Also called as "non-articulate latex ducts", these ducts are independent units which extend as branched structures for long distances in the plant body. They originates as minute structures, elongate quickly and by repeated branching ramify in all directions but do not fuse together. Thus a network is not formed as in latex vessels.

Latex Vessel

Also called "articulate latex ducts", these ducts or vessels are the result of anastamosing of many cells together. They grow more or less as parallel ducts which by means of branching and frequent anastamose form a complex network. Latex vessels are commonly found in many angiosperm families Papaveraceae, Compositae, Euphorbiaceae, Moraceae, etc.

Function

The function of laticiferous ducts is not clearly understood. They may also act as food storage organs or as reservoir of waste products. They may also act as translocatory tissues.

Glandular Tissues

This tissue consists of special structures; the glands. These glands contain some secretory or excretory products. A gland may consist of isolated cells or small group cells with or without a central cavity. They are of various kinds and may be internal or external.

Internal glands are

- Oil-gland secreting essential oils, as in the fruits and leaves of orange, lemon.
- Mucilage secreting glands, as in the betel leaf

- Glands secreting gum, resin, tannin, etc
- Digestive glands secreting enzymes or digestive agents
- Special water secreting glands at the tip of veins

External glands are commonly short hairs tipped by glands. They are

- water-secreting hairs or glands,
- Glandular hairs secreting gum like substances as in tobacco, plumbago, etc
- Glandular hairs secreting irritating, poisonous substances,as in nettles
- Honey glands, as in carnivorous plants.

Hydathode

A hydathode is a type of secretory tissue in leaves, usually found in Angiosperms, that secretes water through pores in the epidermis or margin of leaves, typically at the tip of a marginal tooth or serration. Hydathodes are mainly found in aquatic plants and in some herbaceous plants growing in moist places. They probably evolved from modified stomata, and are connected to the plant vascular system by a vascular bundle. Hydathodes are commonly seen in water lettuce, water hyacinth, rose, balsam, and many other species.

Hydathodes are made of a group of living cells with numerous intercellular spaces filled with water, but few or no chloroplasts, and represent modified bundle-ends. These cells (called epithem cells) open out into one or more sub-epidermal chambers. These, in turn, communicate with the exterior through an open water stomata or open pore. The water stoma structurally resembles an ordinary stoma, but is usually larger and has lost the power of movement.

Hydathodes are involved in the process of guttation, in which positive xylem pressure (due to root pressure) causes liquid to exude from the pores. Some halophytes possess glandular trichomes that actively secrete salt in order to reduce the concentration of cytotoxic inorganic ions in their cytoplasm; this may lead to the formation of a white powdery substance on the surface of the leaf.

Gibberellin

Gibberellins (GAs) are plant hormones that regulate growth and influence various developmental processes, including stem elongation, germination, dormancy, flowering, sex expression, enzyme induction, and leaf and fruit senescence.

Gibberellin was first recognized in 1926 by a Japanese scientist, Eiichi Kurosawa, studying *bakanae*, the "foolish seedling" disease in rice. It was first isolated in 1935 by Teijiro Yabuta and Sumuki, from fungal strains (*Gibberella fujikuroi*) provided by Kurosawa. Yabuta named the isolate as gibberellin.

Interest in gibberellins outside of Japan began after World War II. In the United States, the first research was undertaken by a unit at Camp Detrick in Maryland, via studying seedlings of the bean *Vicia faba*. In the United Kingdom, work on isolating new types of gibberellin was undertaken at Imperial Chemical Industries. Interest in gibberellins spread around the world as the potential for its use on various commercially important plants became more obvious. For example, research that started at the University of California, Davis in the mid-1960s led to its commercial use on Thompson seedless table grapes throughout California by 1962. A known antagonist to gibberellin is paclobutrazol (PBZ), which in turn inhibits growth and induces early fruitset as well as seedset.

Chemistry

***Figure:** Effect of gibberellic acid on cannabis sprout*

All known gibberellins are diterpenoid acids that are synthesized by the terpenoid pathway in plastids and then modified in the endoplasmic reticulum and cytosol until they reach their biologically-active form. All gibberellins are derived via the *ent*-gibberellane skeleton, but are synthesised via *ent*-kaurene. The gibberellins are named GA1 through GAn in order of discovery. Gibberellic acid, which was the first gibberellin to be structurally characterized, is GA3.

As of 2003, there were 126 GAs identified from plants, fungi, and bacteria.

Gibberellins are tetracyclic diterpene acids. There are two classes based on the presence of either 19 or 20 carbons. The 19-carbon gibberellins, such as gibberellic acid, have lost carbon 20 and, in place, possess a five-member lactone bridge that links carbons 4 and 10. The 19-carbon forms are, in general, the biologically active forms of gibberellins. Hydroxylation also has a great effect on the biological activity of the gibberellin. In general, the most biologically active compounds are dihydroxylated gibberellins, which possess hydroxyl groups on both carbon 3 and carbon 13. Gibberellic acid is a dihydroxylated gibberellin.

Bioactive Gas

The bioactive GAs are GA1, GA3, GA4, and GA7. There are three common structural traits between these GAs: 1) a hydroxyl group on C-3β, 2) a carboxyl group on C-6, and 3) a lactone between C-4 and C-10. The 3β-hydroxyl group can be exchanged for other functional groups at C-2 and/or C-3 positions. GA5 and GA6 are examples of bioactive GAs that do not have a hydroxyl group on C-3β. The presence of GA1 in various plant species suggests that it is a common bioactive GA.

Biological Function

Gibberellins are involved in the natural process of breaking dormancy and various other aspects of germination. Before the photosynthetic apparatus develops sufficiently in the early stages of germination, the stored energy reserves of starch nourish the seedling. Usually in germination, the breakdown of starch to glucose in the endosperm begins shortly after the seed is exposed to water. Gibberellins in the seed embryo are believed to signal starch hydrolysis through inducing the synthesis of the enzyme α-amylase in the aleurone cells. In the model for gibberellin-induced production of α-amylase, it is demonstrated that gibberellins (denoted by GA) produced in the scutellum diffuse to the aleurone cells, where they stimulate the secretion α-amylase. α-Amylase then hydrolyses starch, which is abundant in many seeds, into glucose that can be used in cellular respiration to produce energy for the seed embryo. Studies of this process have indicated gibberellins cause higher levels of transcription of the gene coding for the α-amylase enzyme, to stimulate the synthesis of α-amylase.

Gibberellins are produced in greater mass when the plant is exposed to cold temperatures. They stimulate cell elongation, breaking and budding, seedless fruits, and seed germination. They do the last by breaking the seed's dormancy and acting as a chemical messenger. Its hormone binds to a receptor, and Ca2+ activates the protein calmodulin, and the complex binds to DNA, producing an enzyme to stimulate growth in the embryo. A major effect of gibberellins is the degradation of DELLA proteins, the absence of which then allows phytochrome interacting factors to bind to gene promoters and regulate gene expression. Gibberellins are thought to cause DELLAs to become polyubiquitinated and, thus, destroyed by the 26S proteasome pathway.

Biosynthesis

GAs are usually produced from the methylerythritol phosphate (MEP) pathway in higher plants. In this pathway, bioactive GA is produced from trans-geranylgeranyl diphosphate (GGDP). In the MEP pathway, three classes of enzymes are used to yield GA from GGDP: 1) terpene synthases (TPSs), 2) cytochrome P450 monooxygenases (P450s), and 3) 2-oxoglutarate–dependent dioxygenases (2ODDs). There are 8 steps in the methylerythritol phosphate pathway: - 1) GGDP is converted to ent-copalyl diphosphate (ent-CPD) by ent-copalyl diphosphate synthase - 2) etn-CDP is converted to ent-kaurene by ent-kaurene synthase - 3) ent-kaurene is converted to ent-kaurenol by ent-kaurene oxidase (KO) - 4) ent-kaurenol is converted to ent-kaurenal by KO - 5) ent-kaurenal is converted to ent-kaurenoic acid by KO - 6) ent-kaurenoic acid is converted to ent-7a-hydroxykaurenoic acid by ent-kaurene acid oxidase (KAO) - 7) ent-7a-hydroxykaurenoic acid is converted to GA12-aldehyde by KAO - 8) GA12-aldehyde is converted to GA12 by KAO. GA12 is processed to the bioactive GA4 by oxidations on C-20 and C-3, which is accomplished by 2 soluble ODDs: GA 20-oxidase and GA 3-oxidase.

Sites of Biosynthesis

Most bioactive GAs are located in actively growing organs on plants. Both GA20ox and GA3ox genes (genes coding for GA 20-oxidase and GA 3-oxidase) and the SLENDER1 gene (a GA signal transduction gene) are found in growing organs on rice, which suggests bioactive GA synthesis occurs at their site of action in growing organs in plants. During flower development, the tapetum of anthers is believed to a primary site of GA biosynthesis.

Differences between Biosynthesis in Fungi and Lower Plants

Arabidopsis, a plant, and Gibberella fujikuroi, a fungus, possess different GA pathways and enzymes. P450s in fungi perform functions analogous to the functions of KAOs in plants. The function of CPS and KS in plants is performed by a single enzyme, CPS/KS, in fungi. In fungi, the GA biosynthesis genes are found on one chromosome, but in plants, they are found randomly on multiple chromosomes.

Gibberellin Metabolism Genes

One or two genes encode the enzymes responsible for the first steps of GA biosynthesis in Arabidopsis and rice. The null alleles of the genes encoding CPS, KS, and KO result in GA-deficient Arabidopsis dwarves. Multigene families encode the 2ODDs that catalyze the formation of GA12 to bioactive GA4.

AtGA3ox1 and AtGA3ox2, two of the four genes that encode GA3ox in Arabidopsis, affect vegetative development. Environmental stimuli regulate AtGA3ox1 and AtGA3ox2 activity during seed germination. In Arabidopsis, GA20ox overexpression leads to an increase in GA concentration.

Deactivation

Several mechanisms for inactivating GAs have been identified. 2β-hydroxylation deactivates GA, and is catalyzed by GA2-oxidases (GA2oxs). Some GA2oxs use C19-GAs as substrates, and other GA2oxs use C20-GAs.

Cytochrome P450 mono-oxygenase, encoded by elongated uppermost internode (eui), converts GAs into 16α,17-epoxides. Rice eui mutants amass bioactive GAs at high levels, which suggests cytochrome P450 mono-oxygenase is a main enzyme responsible for deactivation GA in rice.

The Gamt1 and gamt2 genes encode enzymes that methylate the C-6 carboxyl group of GAs. In a gamt1 and gamt2 mutant, concentrations of GA is developing seeds is increased.

Regulation

Regulation by Other Hormones

The auxin indole-3-acetic acid (IAA) regulates concentration of GA1 in elongating internodes in peas. Removal of IAA by removal of the apical bud, the auxin source, reduces the concentration of GA1, and reintroduction of IAA reverses these effects to increase the concentration of GA1. This phenomenon has also been observed in tobacco plants. Auxin increases GA 3-oxidation and decreases GA 2-oxidation in barley. Auxin also regulates GA biosynthesis during fruit development in peas. These discoveries in different plant species suggest the auxin regulation of GA metabolism may be a universal mechanism.

Ethylene decreasing the concentration of bioactive GAs.

Regulation by Environmental Factors

Recent evidence suggests fluctuations in GA concentration influence light-regulated seed germination, photomorphogenesis during de-etiolation, and photoperiod regulation of stem elongation and flowering. Microarray analysis showed about one fourth cold-responsive genes are related to GA-regulated genes, which suggests GA influences response to cold temperatures. Plants reduce growth rate when exposed to stress. A relationship between GA levels and amount of stress experienced has been suggested in barley.

Role in Seed Development

Bioactive GAs and abcisic acid levels have an inverse relationship and regulate seed development and germination. Levels of FUS3, an Arabidopsis transcription factor, are upregulated by ABA and downregulated by GA, which suggests that there is a regulation loop that establishes the balance of GA and ABA.

Gibberellin Homeostasis

Feedback and feedforward regulation maintains the levels of bioactive GAs in plants. Levels of AtGA20ox1 and AtGA3ox1 expression are increased in a GA deficient environment, and decreased after the addition of bioactive GAs, Conversely, expression of AtGA2ox1 and AtGA2ox2, GA deactivation genes, is increased with addition of GA.

Impact on the "Green Revolution"

A chronic food shortage was feared during the rapid climb in world population in the 1960s. This was averted with the development of a high-yielding variety of rice. This variety of semi-dwarf rice is called IR8, and it has a short height because of a mutation in the sd1 gene. Sd1 encodes GA20ox, so a mutant sd1 is expected to exhibit a short height that is consistent with GA deficiency.

Receptors

GA receptors must have four traits: 1) reversibly bind GA, 2) GA saturability 3) high affinity for bioactive GA, and 4) reasonable ligand specificity for bioactive GAs.

Cereal Aleurone Cells

The first reported GA-binding protein (GBP) activity was discovered in wheat seed aleurone homogenates by using the GA-dependent induction of aleuron hydrolytic enzymes. Alpha-amylase induction in aleurone protoplasts is dependent on GA levels, but induction can still occur if GA does not pass through the plasma membrane. The existence of GA perception outside the cell and a plasma membrane GA receptor in aleurone cells can be concluded from these observations. Trimeric G proteins are involved in GA signaling, which supports the previous theory for plasma membrane GA receptors. Four different GBPs have been discovered in the plasma membrane of aleuron cells. Cereal aleurone cells could be unique in GA perception because they are some of the few plant cells that cannot synthesize bioactive GA.

GID1 Receptor

The rice GID1 gene encodes an unknown protein that only has a high affinity for bioactive GAs. In yeast, GID1 binds SLR1, GA signaling repressor, depending on GA levels. GID1 is a soluble GA receptor mediating GA signaling. The GA-GID1 complex interacts with SLR1 to transduce the GA signal to SLR1. SLR1 increased the GA-binging activity of GID1 threefold. Arabidopsis has three GID1 homologs that bind GA and interact with DELLAs: AtGID1a, b, and c.

Gibberellin and DELLA Proteins

DELLA Proteins: Repressors of GA-dependent Processes

DELLA proteins act as intracellular repressors of GA responses, DELLAs inhibit seed germination, seed growth, and other GA-dependent pathways, but GA can reverse these effects.

GA-GID1-DELLA Complex

For GA to bind to the GID1 receptor, the C3-hydroxyl on GA must form a hydrogen bond to the tyrosine 31 residue of GID1, which induces a conformational change in GID1 to enclose GA. Once GA is trapped in the GID1 pocket, the lid of the pocket binds to DELLA to form the GA-GID1-DELLA complex.

GA Promotes Proteasome-dependent Degradation of DELLAs

In the absence of GA, many DELLAs are present to repress GA responses, but the formation of the GA-GID1-DELLA complex increases the degradation of DELLAs. The SCF complex is composed of SKP1, CULLIN, and F-BOX proteins. F-box proteins catalyze the formation of polyubiquitin on target proteins to be degraded by the 26S proteasome. The formation of the GA-GID1-DELLA complex is believed to cause a conformational change in DELLA, which enhances recognition of DELLAs by F-box proteins. Next, the SCF complex promotes ubiquitylation of DELLAs, which are then degraded by the 26S proteasome. Degradation of DELLAs allows GA regulated growth to resume. Thus, GA stimulates growth by activating the degradation of DELLAs.

Allelopathic Effects of Dominant Plants in Forest

In an ecological community, the species which is most numerous and forms the bulk of biomass may be considered the dominant species. Ecological dominance may also be defined as the species that has the most influence on other species in the same environment. Dominant species include plants and animals that influence the ecological conditions of an environment by their size, abundance, or behaviour and determine which other animals or plants can survive in that environment. In some environments, there may be one or more dominant species.

Good examples of ecological domination in the plant world are the forest communities of the Rocky Mountains. After a forest fire, plants and trees progress through various stages, with small plants like grasses and ferns growing back first. Eventually, small trees like aspen and birch take root and sprout upward cutting off sunlight from those smaller ground plants on the forest floor. After a number of years, coniferous trees like pines and spruces will grow above the smaller trees. At each stage of growth, one species of plant succeeds the previous species and, for a time, exists as the dominant species in the forest ecosystem.

In the animal kingdom, the top predator may become the dominant species. A good example is the Yellowstone Lake ecosystem in Wyoming, where predatory lake trout were illegally introduced in 1994. Prior to the introduction of the piscivorous, or fish eating, lake trout, the Yellowstone cutthroat trout was the dominant species in the lake ecosystem. In 2011, if not controlled, lake trout could reduce the Yellowstone cutthroat trout population by as much as 90% over the next 20 years and might eliminate them completely. The lake trout would then be considered the dominant species in this ecological community.

The top predator might not always be the dominant species. Sometimes a species attains dominance through sheer numbers, and in this case, the total biomass of a species makes it the dominant species. Large numbers of a single species can exert tremendous influence on an ecosystem. One simple example of this would be a locust infestation in a wheat field. With an abundant food supply and no predators, a population of locusts can increase its numbers over a very short period of time, becoming the dominant species in this ecosystem.

Ecological dominance is the degree to which a taxon is more numerous than its competitors in an ecological community, or makes up more of the biomass. Most ecological communities are defined by their dominant species.

- In many examples of wet woodland in western Europe, the dominant tree is alder (*Alnus glutinosa*).

- In temperate bogs, the dominant vegetation is usually species of *Sphagnum* moss.
- Tidal swamps in the tropics are usually dominated by species of mangrove (*Rhizophoraceae*)
- Some sea floor communities are dominated by brittle stars.
- Exposed rocky shorelines are dominated by sessile organisms such as barnacles and limpets.

Dominant plant species with high abundance and biomass are major contributors to nutrient and energy pools in many ecosystems. Therefore, the loss of dominant plant species can result in a reduction of ecosystem function, particularly if the remaining species in the community do not have similar functional roles. Many efforts have been made to document the effect of dominant species loss. These include observational data from communities where the dominant species has declined, comparative studies of mono- vs. polyculture assemblages, as well as synthetic assemblage and removal experiments.

Apical Dominance

Apical dominance is the phenomenon whereby the main central stem of the plant is dominant over (i.e., grows more strongly than) other side stems; on a branch the main stem of the branch is further dominant over its own side branchlets.

Plant physiology describes apical dominance as the control exerted by the terminal bud (and shoot apex) over the outgrowth of lateral buds.

Apical dominance occurs when the shoot apex inhibits the growth of lateral buds so that the plant may grow vertically. It is important for the plant to devote energy to growing upward so that it can get more light to undergo photosynthesis. If the plant utilizes available energy for growing upward, it may be able to outcompete other individuals in the near vicinity. Plants that were capable of outcompeting neighbouring plants likely had higher fitness. Apical dominance is therefore most likely adaptive.

Typically, the end of a shoot contains an apical bud, which is the location where shoot growth occurs. The apical bud produces an auxin (IAA) that inhibits growth of the lateral buds further down on the stem towards the axillary bud. It was first discovered that the plant hormone auxin likely regulates apical dominance in 1934. Auxin is predominantly produced in the growing shoot apex and is transported throughout the plant via the xylem and diffuses into lateral buds which prevents elongation.

When the apical bud is removed, the lowered IAA concentration allows the lateral buds to grow and produce new shoots, which compete to become the lead growth.

Figure: *Weeping larch showing growth habit is lacking apical dominance.*

Apex Removal

Plant physiologists have identified four different stages the plant goes through after the apex is removed (Stages I-IV). The four stages are referred to as

1. lateral bud formation,
2. "imposition of inhibition" (apical dominance),
3. initiation of lateral bud outgrowth following decapitation, and
4. elongation and development of the lateral bud into a branch.

These stages can also be defined by the hormones that are regulating the process which are as follows: Stage I, cytokinin promoted, causing the lateral bud to form since cytokinin plays a role in cell division; Stage II, auxin is promoted, resulting in apical dominance ("imposition of inhibition"); Stage III, cytokinin released resulting in outward growth of the lateral bud; and Stage IV, auxin is decreased and gibberellic acid is promoted which results in cell division, enabling the bud or branch to continue outward growth.

More simply stated, lateral bud formation is inhibited by the shoot apical meristem (SAM). The lateral bud primordium (from which the lateral bud develops) is located below SAM. The shoot tip rising from the SAM inhibits the growth of the lateral bud by repressing auxin. When the shoot is cut off, the lateral bud begins to lengthen which is mediated by a release of cytokinin. Once the apical dominance has been lifted from the plant, elongation and lateral growth is promoted and the lateral buds grow into new branches. When lateral bud formation prevents the plant from growing upward, it is undergoing lateral dominance. Often, lateral dominance can be triggered by decapitating the SAM or artificially decreasing the concentration of auxin in plant tissues.

Applications

When the apical bud is removed, the lowered IAA concentration allows the lateral buds to grow and produce new shoots, which compete to become the lead growth. Pruning techniques such as coppicing and pollarding make use of this natural response to damage to direct plant growth and produce a desired shape, size, and/or productivity level for the plant. The principle of apical dominance is manipulated for espalier creation, hedge building, or artistic sculptures called topiary. If the SAM is removed, it stimulates growth in the lateral direction. By careful pruning, it is possible to create remarkable designs or patterns.

Some fruit trees have strong apical dominance, and young trees can become "leggy", with poor side limb development. Apical dominance can be reduced in this case, or in cases where limbs are broken off by accident, by cutting off the auxin flow above side buds that one wishes to stimulate. This is often done by orchardists for young trees.

Occasionally, strong apical dominance is advantageous, as in the "Ballerina" apple trees. These trees are intended to be grown in small gardens, and their strong apical dominance combined with a dwarfing rootstock gives a compact narrow tree with very short fruiting side branches.

Enzymes Bioassays

Bioassay (commonly used shorthand for biological assay), or biological standardization is a type of scientific experiment. A bioassay involves the use of a live animal (*in vivo*) or tissue (*in vitro*) to determine the biological activity of a substance, such as a hormone or drug. Bioassays are typically conducted to measure the effects of a substance on a living organism and are essential in the development of new drugs and in monitoring environmental pollutants. Both are procedures by which the potency or the nature of a substance is estimated by studying its effects on living matter. A bioassay can also be used to determine the concentration of a particular constitution of a mixture.

Use

Bioassays are procedures that can determine the concentration or purity or biological activity of a substance such as vitamin, hormone or plant growth factor by measuring the effect on an organism, tissue, cells, enzyme or receptor. Bioassays may be qualitative or quantitative. Qualitative bioassays are used for assessing the physical effects of a substance that may not be quantified, such as abnormal development or deformity. An example of a qualitative bioassay includes Arnold Adolph Berthold's famous experiment on castrated chickens.

This analysis found that by removing the testes of a chicken, it would not develop into a rooster because the endocrine signals necessary for this process were not available. Quantitative bioassays involve estimation of the concentration or potency of a substance by measurement of the biological response that it produces. Quantitative bioassays are typically analyzed using the methods of biostatistics.

Definition

"The determination of the relative strength of a substance (e.g., a drug or hormone) by comparing its effect on a test organism with that of a standard preparation." ....

Purpose

1. Measurement of the pharmacological activity of new or chemically undefined substances
2. Investigation of the function of endogenous mediators
3. Determination of the side-effect profile, including the degree of drug toxicity
4. Measurement of the concentration of known substances (alternatives to the use of whole animals have made this use obsolete)
5. Assessing the amount of pollutants being released by a particular source, such as wastewater or urban runoff.
6. Determining the specificity of certain enzymes to certain substrates.

Types

Bioassays are of two types:

Quantal

A quantal assay involves an "all or none response". For example: Insulin induced hypoglycemic convulsive reaction or the cardiac arrest caused by digitalis. The response is either +ve or -ve, there is no intermediate response e.g.—either convulsion occurs or doesn't occur; similarly is with cardiac arrest.

In case of toxicity studies, the animal receiving a dose of drug either dies or does not die. Also, no intermediate response is possible. This is also known as the "all or none" response assay. The quantal method though not precise is employed for bioassay of substance in the following ways:

(a) Comparison of threshold response or

(b) Comparison of effective dose (ED_{50}) or median lethal dose (LD_{50})

Graded

Graded assays are based on the observation that there is a proportionate increase in the observed response following an increase in the concentration or dose. The parametres employed in such bioassays are based on the nature of the effect the substance is expected to produce. For example: contraction of smooth muscle preparation for assaying histamine or the study of blood pressure response in case of adrenaline.

A graded bioassay can be performed by employing any of the below-mentioned techniques. The choice of procedure depends on:

1. the precision of the assay required
2. the quantity of the sample substance available
3. the availability of the experimental animals.

Techniques

1. Matching Bioassay
2. Interpolation Method
3. Bracketing Method
4. Multiple Point Bioassay (i.e.-Three-point, Four-point and Six Point Bioassay)

Matching Bioassay: It is the simplest type of the bioassay. In this type of bioassay, response of the test substance taken first and the observed response is tried to match with the standard response. Several responses of the standard drug are recorded till a close matching point to that of the test substance is observed. A corresponding concentration is thus calculated. This assay is applied when the sample size is too small. Since the assay does not involve the recording of concentration response curve, the sensitivity of the preparation is not taken into consideration. Therefore, precision and reliability is not very good.

Interpolation bioassay: Bioassays are conducted by determining the amount of preparation of unknown potency required to produce a definite effect on suitable test animals or organs or tissue under standard conditions. This effect is compared with that of a standard. Thus the amount of the test substance required to produce the same biological effect as a given quantity the unit of a standard preparation is compared and the potency of the unknown is expressed as a % of that of the standard by employing a simple formula.

Many times, a reliable result cannot be obtained using this calculation. Therefore it may be necessary to adopt more precise methods of calculating potency based upon observations of relative, but not necessarily equal effects, likewise, statistical methods may also be employed. The data (obtained from either of assay techniques used) on which bioassay are based may be classified as quantal or graded response. Both these depend ultimately on plotting or making assumption concerning the form of DRC.

Environmental Bioassays

Environmental bioassays are generally a broad-range survey of toxicity. A toxicity identification evaluation is conducted to determine what the relevant toxicants are. Although bioassays are beneficial in determining the biological activity within an organism, they can often be time-consuming and laborious. Organism-specific factors may result in data that is not applicable to others in that species. For these reasons, other biological techniques are often employed, including radioimmunoassays.

Water pollution control requirements in the United States require some industrial dischargers and municipal sewage treatment plants to conduct bioassays. These procedures, called whole effluent toxicity tests, include acute toxicity tests as well as chronic test methods. The methods involve exposing living aquatic organisms to samples of wastewater.

RuBisCO

Ribulose-1,5-bisphosphate carboxylase/oxygenase, commonly known by the abbreviation RuBisCO, is an enzyme involved in the first major step of carbon fixation, a process by which atmospheric carbon dioxide is converted by plants to energy-rich molecules such as glucose. In chemical terms, it catalyzes the carboxylation of ribulose-1,5-bisphosphate (also known as *RuBP*). It is probably the most abundant protein on Earth.

RuBisCO vs Alternative Carbon Fixation Pathways

RuBisCO is important biologically because it catalyzes the primary chemical reaction by which inorganic carbon enters the biosphere. While many autotrophic bacteria and archaea fix carbon via the reductive acetyl CoA pathway, the 3-hydroxypropionate cycle, or the reverse Krebs cycle, these pathways are relatively smaller contributors to global carbon fixation than that catalyzed by RuBisCO. Phosphoenolpyruvate carboxylase, unlike RuBisCO, only temporarily fixes carbon. Reflecting its importance, RuBisCO is the most abundant protein in leaves, accounting for 50% of soluble leaf protein in C_3 plants (20–30% of total leaf nitrogen) and 30% of soluble leaf protein in C_4 plants (5–9% of total leaf nitrogen). Given its important role in the biosphere, the genetic engineering of RuBisCO in crops is of continuing interest.

Structure

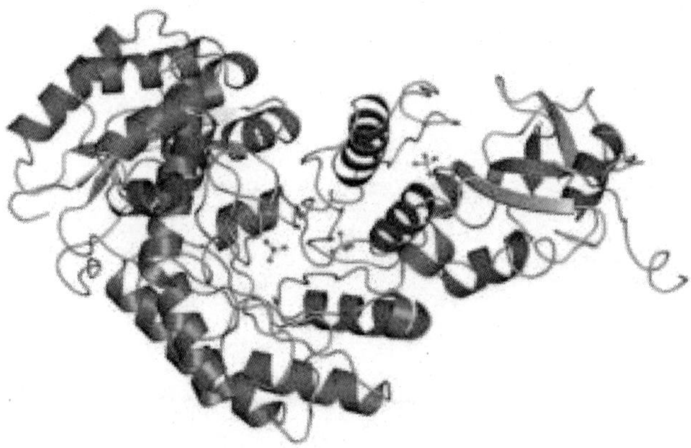

Figure: *PDB structure 3rub: the ribulose-1,5-bisphosphate carboxylase/oxygenase from tobacco.*

In plants, algae, cyanobacteria, and phototrophic and chemoautotrophic proteobacteria, the enzyme usually consists of two types of protein subunit, called the large chain (L, about 55,000 Da) and the small chain (S, about 13,000 Da). The *large-chain* gene (*rbcL*) is part of the chloroplast DNA molecule in plants. There are typically several related *small-chain* genes in the nucleus of plant cells, and the small chains are imported to the stromal compartment of chloroplasts from the cytosol by crossing the outer chloroplast membrane. The enzymatically active substrate (ribulose 1,5-bisphosphate) binding sites are located in the large chains that form dimers as shown in Figure 1 (above, right) in which amino acids from each large chain contribute to the binding sites. A total of eight large-chains (= 4 dimers) and eight small chains assemble into a larger complex of about 540,000 Da. In some proteobacteria and dinoflagellates, enzymes consisting of only large subunits have been found.

Magnesium ions (Mg_{2+}) are needed for enzymatic activity. Correct positioning of Mg_{2+} in the active site of the enzyme involves addition of an "activating" carbon dioxide molecule (CO_2) to a lysine in the active site (forming a carbamate). Formation of the carbamate is favoured by an alkaline pH. The pH and the concentration of magnesium ions in the fluid compartment (in plants, the stroma of the chloroplast) increases in the light. The role of changing pH and magnesium ion levels in the regulation of RuBisCO enzyme activity is discussed below.

Enzymatic Activity

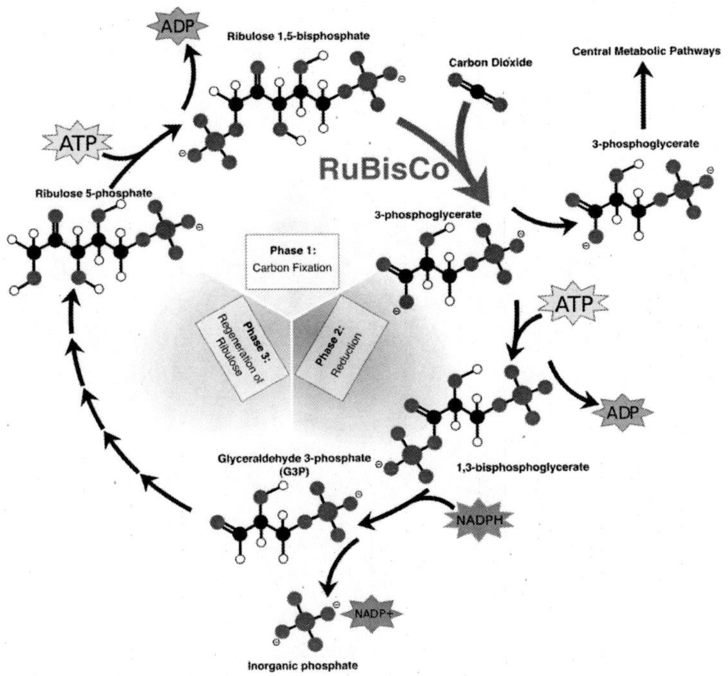

Figure: *Overview of the Calvin cycle and carbon fixation*

As shown in figure (left), RuBisCO is one of many enzymes in the Calvin cycle.

Substrates

During carbon fixation, the substrate molecules for RuBisCO are ribulose-1,5-bisphosphate, carbon dioxide (distinct from the "activating" carbon dioxide). RuBisCO also catalyses a reaction between ribulose-1,5-bisphosphate and molecular oxygen (O_2) instead of carbon dioxide (CO_2).

Products

When carbon dioxide is the substrate, the product of the carboxylase reaction is a highly unstable six-carbon phosphorylated intermediate known as 3-keto-2-carboxyarabinitol-1,5-bisphosphate, which decays virtually instantaneously into two molecules of glycerate-3-phosphate.

The extremely unstable molecule created by the initial carboxylation was unknown until 1988, when it was isolated. The 3-phosphoglycerate can be used to produce larger molecules such as glucose. When molecular oxygen is the substrate, the products of the oxygenase reaction are phosphoglycolate and 3-phosphoglycerate.

Phosphoglycolate is recycled through a sequence of reactions called photorespiration, which involves enzymes and cytochromes located in the mitochondria and peroxisomes. In this process, two molecules of phosphoglycolate are converted to one molecule of carbon dioxide and one molecule of 3-phosphoglycerate, which can reenter the Calvin cycle. Some of the phosphoglycolate entering this pathway can be retained by plants to produce other molecules such as glycine. At ambient levels of carbon dioxide and oxygen, the ratio of the reactions is about 4 to 1, which results in a net carbon dioxide fixation of only 3.5. Thus, the inability of the enzyme to prevent the reaction with oxygen greatly reduces the photosynthetic capacity of many plants. Some plants, many algae, and photosynthetic bacteria have overcome this limitation by devising means to increase the concentration of carbon dioxide around the enzyme, including C_4 carbon fixation, crassulacean acid metabolism, and the use of pyrenoid.

Rate of Enzymatic Activity

Some enzymes can carry out thousands of chemical reactions each second. However, RuBisCO is slow, being able to fix only 3-10 carbon dioxide molecules each second per molecule of enzyme. The reaction catalyzed by RuBisCO is, thus, the primary rate-limiting factor of the Calvin cycle during the day. Nevertheless, under most conditions, and when light is not otherwise limiting photosynthesis, the speed of RuBisCO responds positively to increasing carbon dioxide concentration.

However, our descriptive knowledge will become more usable when we can translate them into quantitative models that can enable us to calculate the outcome of the reaction under a given condition. Since RubisCO reacts with RuBP (ribulose 1,5 bisphosphate) first to produces enediol and next with CO_2 that after some intermediate changes produces PGA (3-phosphoglycerate), a biochemical model is developed to represent the effects of these steps quantitatively.

Since carboxylation or fixation of CO2 is possible only after the synthesis of enediol, thus it is suggested that the role of RubisCO is to produce enediol that is carboxylase and oxygenase (EnCO). Accordingly, RubisCO is called enolase-phosphglycerase (EPGase) since it is neither carboxylase nor oxygenase.

Regulation of its Enzymatic Activity

RuBisCO is usually only active during the day as ribulose 1,5-bisphosphate is not regenerated in the dark. This is due to the regulation of several other enzymes in the Calvin cycle. In addition, the activity of RuBisCO is coordinated with that of the other enzymes of the Calvin cycle in several ways.

Regulation by Ions

Upon illumination of the chloroplasts, the pH of the stroma rises from 7.0 to 8.0 because of the proton (hydrogen ion, H+) gradient created across the thylakoid membrane. At the same time, magnesium ions (Mg2+) move out of the thylakoids, increasing the concentration of magnesium in the stroma of the chloroplasts. RuBisCO has a high optimal pH (can be >9.0, depending on the magnesium ion concentration) and, thus, becomes "activated" by the addition of carbon dioxide and magnesium to the active sites as described above.

Regulation by RuBisCO Activase

In plants and some algae, another enzyme, RuBisCO activase, is required to allow the rapid formation of the critical carbamate in the active site of RuBisCO. RuBisCO activase is required because the ribulose 1,5-bisphosphate (RuBP) substrate binds more strongly to the active sites lacking the carbamate and markedly slows down the "activation" process. In the light, RuBisCO activase promotes the release of the inhibitory, or — in some views — storage RuBP from the catalytic sites. Activase is also required in some plants (e.g., tobacco and many beans) because, in darkness, RuBisCO is inhibited (or protected from hydrolysis) by a competitive inhibitor synthesized by these plants, a substrate analog 2-Carboxy-D-arabitinol 1-phosphate (CA1P). CA1P binds tightly to the active site of carbamylated RuBisCO and inhibits catalytic activity.

In the light, RuBisCO activase also promotes the release of CA1P from the catalytic sites. After the CA1P is released from RuBisCO, it is rapidly converted to a non-inhibitory form by a light-activated CA1P-phosphatase. Finally, once every several hundred reactions, the normal reactions with carbon dioxide or oxygen are not completed, and other inhibitory substrate analogs are formed in the active site. Once again, RuBisCO activase can promote the release of these analogs from the catalytic sites and maintain the enzyme in a catalytically active form. The properties of activase limit the photosynthetic potential of plants at high temperatures. CA1P has also been shown to keep RuBisCO in a conformation that is protected from proteolysis. At high temperatures,

RuBisCO activase aggregates and can no longer activate RuBisCO. This contributes to the decreased carboxylating capacity observed during heat stress.

Regulation by ATP/ADP and Stromal Reduction/Oxidation State through the Activase

The removal of the inhibitory RuBP, CA1P, and the other inhibitory substrate analogs by activase requires the consumption of ATP. This reaction is inhibited by the presence of ADP, and, thus, activase activity depends on the ratio of these compounds in the chloroplast stroma. Furthermore, in most plants, the sensitivity of activase to the ratio of ATP/ADP is modified by the stromal reduction/oxidation (redox) state through another small regulatory protein, thioredoxin. In this manner, the activity of activase and the activation state of RuBisCO can be modulated in response to light intensity and, thus, the rate of formation of the ribulose 1,5-bisphosphate substrate.

Regulation by Phosphate

In cyanobacteria, inorganic phosphate (P_i) participates in the co-ordinated regulation of photosynthesis. P_i binds to the RuBisCO active site and to another site on the large chain where it can influence transitions between activated and less active conformations of the enzyme. Activation of bacterial RuBisCO might be particularly sensitive to P_i levels, which can act in the same way as RuBisCO activase in higher plants.

Regulation by Carbon Dioxide

Since carbon dioxide and oxygen compete at the active site of RuBisCO, carbon fixation by RuBisCO can be enhanced by increasing the carbon dioxide level in the compartment containing RuBisCO (chloroplast stroma). Several times during the evolution of plants, mechanisms have evolved for increasing the level of carbon dioxide in the stroma. The use of oxygen as a substrate appears to be a puzzling process, since it seems to throw away captured energy. However, it may be a mechanism for preventing overload during periods of high light flux. This weakness in the enzyme is the cause of photorespiration, such that healthy leaves in bright light may have zero net carbon fixation when the ratio of O2 to CO2 reaches a threshold at which oxygen is fixed instead of carbon. This phenomenon is primarily temperature-dependent. High temperature decreases the concentration of CO2 dissolved in the moisture in the leaf tissues. This phenomenon is also related to water stress. Since plant leaves are evaporatively cooled, limited water causes high leaf temperatures. C_4 plants use the enzyme

PEP carboxylase initially, which has a higher affinity for CO2. The process first makes a 4-carbon intermediate compound, which is shuttled into a site of C_3 photosynthesis then de-carboxylated, releasing CO2 to boost the concentration of CO2, hence the name C_4 plants.

Crassulacean acid metabolism (CAM) plants keep their stomata (on the underside of the leaf) closed during the day, which conserves water but prevents the light-independent reactions (a.k.a the Calvin Cycle) from taking place, since these reactions require CO_2 to pass by gas exchange through these openings. Evaporation through the upper side of a leaf is prevented by a layer of wax.

Genetic Engineering

Since RuBisCO is often rate-limiting for photosynthesis in plants, it may be possible to improve photosynthetic efficiency by modifying RuBisCO genes in plants to increase its catalytic activity and/or decrease the rate of the oxygenation activity. This could improve biosequestration of CO_2 and be an important climate change strategy. Approaches that have begun to be investigated include expressing RuBisCO genes from one organism in another organism, increasing the level of expression of RuBisCO subunits, expressing RuBisCO small chains from the chloroplast DNA, and altering RuBisCO genes so as to try to increase specificity for carbon dioxide or otherwise increase the rate of carbon fixation.

One avenue is to introduce RuBisCO variants with naturally high specificity values such as the ones from the red alga *Galdieria partita* into plants. This might be expected to improve the photosynthetic efficiency of crop plants although possible negative impacts have yet to be studied. Advances in this area include the replacement of the tobacco enzyme with that of the purple photosynthetic bacterium *Rhodospirillum rubrum*.

A recent theory explores the trade-off between the relative specificity (i.e., ability to favour CO2 fixation over O2 incorporation, which leads to the energy-wasteful process of photorespiration) and the rate at which product is formed. The authors conclude that RuBisCO may actually have evolved to reach a point of 'near-perfection' in many plants (with widely varying substrate availabilities and environmental conditions), reaching a compromise between specificity and rate of reaction.

Since photosynthesis is the single most effective natural regulator of carbon dioxide in the Earth's atmosphere, a biochemical model of RuBisCO reaction is used as the core module of climate change models. Thus, a correct model of this reaction is essential to the basic understanding of the relations and interactions of environmental models. A new theory and model of the biochemical reaction of photosynthesis and the drawbacks of today's most widely used model of photosynthesis is discussed in the new volume of Advances in Photosynthesis and Respiration

Photorespiration

Photorespiration (also known as the oxidative photosynthetic carbon cycle, or C2 photosynthesis) is a process in plant metabolism which attempts to ameliorate the consequences of a wasteful oxygenation reaction by the enzyme RuBisCO. The desired reaction is the addition of carbon dioxide to RuBP (carboxylation), a key step in the Calvin-Benson cycle, however approximately 25% of reactions by RuBisCO instead add oxygen to RuBP (oxygenation), producing a product that cannot be used within the Calvin-Benson cycle. This process reduces efficiency of photosynthesis, potentially reducing photosynthetic output by 25% in C_3 plants. Photorespiration involves a complex network of enzyme reactions that exchange metabolites between chloroplasts, leaf peroxisomes and mitochondria.

The oxygenation reaction of RuBisCO is a wasteful process because 3-Phosphoglycerate is created at a reduced rate and higher metabolic cost compared with RuBP carboxylase activity. While photorespiratory carbon cycling results in the formation of G3P eventually, there is still a net loss of carbon (around 25% of carbon fixed by photosynthesis is re-released as CO_2) and nitrogen, as ammonia. The ammonia must be detoxified at a substantial cost to the cell. Photorespiration also incurs a direct cost of one ATP and one NAD(P)H.

While it is common to refer to the entire process as photorespiration, technically the term refers only to the metabolic network which acts to rescue the products of the oxygenation reaction (phosphoglycolate).

Photorespiratory Reactions

Addition of molecular oxygen to ribulose-1,5-bisphosphate produces 3-phosphoglycerate (PGA) and 2-phosphoglycolate (2PG, or PG). PGA is the normal product of carboxylation, and productively enters the Calvin cycle. Phosphoglycolate, however, inhibits certain enzymes involved in photosynthetic carbon fixation (hence is often said to be an 'inhibitor of photosynthesis'). It is also relatively difficult to recycle: in higher plants it is salvaged by a series of reactions in the peroxisome, mitochondria, and again in the peroxisome where it is converted into glycerate. Glycerate reenters the chloroplast and by the same transporter that exports glycolate. A cost of 1 ATP is associated with conversion to 3-phosphoglycerate (PGA) (Phosphorylation), within the chloroplast, which is then free to re-enter the Calvin cycle.

There are several costs associated with this metabolic pathway; one being the production of hydrogen peroxide in the peroxisome (associated with the conversion of glycolate to glyoxylate). Hydrogen peroxide is a dangerously strong oxidant which must be immediately broken down into water and oxygen by the enzyme catalase. The conversion of 2x 2Carbon glycine to 1 C3 serine in the mitochondria by the enzyme glycine-decarboxylase is a key step, which releases CO_2, NH_3, and reduces NAD to NADH. Thus, 1 CO2 molecule is produced for every 3 molecules of O2 (two deriving from the activity of RuBisCO, the third from peroxisomal oxidations). The assimilation of NH_3 occurs via the GS-GOGAT cycle, at a cost of one ATP and one NADPH.

Cyanobacteria have three possible pathways through which they can metabolise 2-phosphoglycolate. They are unable to grow if all three pathways are knocked out, despite having a carbon concentrating mechanism that should dramatically reduce the rate of photorespiration.

Substrate Specificity of RuBisCO

The oxidative photosynthetic carbon cycle reaction is catalyzed by RuBP oxygenase activity:

RuBP + O_2 → Phosphoglycolate + 3-phosphoglycerate + 2H+

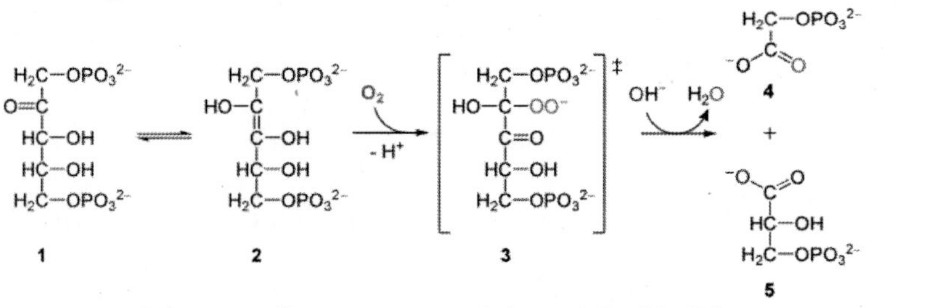

Figure: *Oxygenase activity of RubisCO*

During the catalysis by RuBisCO, an 'activated' intermediate is formed (an enediol intermediate) in the RuBisCO active site. This intermediate is able to react with either CO_2 or O_2. It has been demonstrated that the specific shape of the RuBisCO active site acts to encourage reactions with CO2. Although there is a significant "failure" rate (~25% of reactions are oxygenation rather than carboxylation), this represents significant favouring of CO_2, when the relative abundance of the two gasses is taken into account: in the current atmosphere, O_2 is approximately 500 times more abundant, and in solution O_2 is 25X more abundant than CO_2. The ability of RuBisCO to specify between the two gasses is known as its selectivity factor (or Srel), and it varies between species (although there is not much variation within the vascular plants).

A suggested explanation into Rubisco's inability to discriminate completely between CO_2 and O_2 is that it is an evolutionary relic: The early atmosphere in which primitive plants originated contained very little oxygen, the early evolution of RuBisCO was notinfluenced by its ability to discriminate between O_2 and carbon dioxide.

Conditions which Increase Photorespiration

Photorespiration rates are increased by:

Altered Substrate Availability: Lowered CO2 or Increased O2

Factors which influence this include the atmospheric abundance of the two gases, the supply of the gases to the site of fixation (i.e. in land plants: whether the stomata are open or closed), the length of the liquid phase (how far these gases have to diffuse through water in order to reach the reaction site). For example, when the stomata are closed to prevent water loss during drought: this limits the CO_2 supply, while O_2 production within the leaf will continue. In algae (and plants which photosynthesise underwater) gases have to diffuse significant distances through water, which results in a decrease in the availability of CO_2 relative to O_2. It has been predicted that the increase in ambient CO_2 concentrations predicted over the next 100 years may reduce the rate of photorespiration in most plants by around 50%.

Increased Temperature

At higher temperatures RuBisCO is less able to discriminate between CO_2 and O_2. This is because the enediol intermediate is less stable. Increasing temperatures also reduce the solubility of CO_2, thus reducing the concentration of CO_2 relative to O_2 in the chloroplast.

High Light

Photorespiration may be used as a mechanism to dissipate excess energy at high irradiance levels.

Biological Adaptation to Minimize Photorespiration

Certain species of plants or algae have mechanisms to reduce uptake of molecular oxygen by RuBisCO. These are commonly referred to as Carbon Concentrating Mechanisms (CCMs), as they increase the concentration of CO_2 so that Rubisco is less likely to produce glycolate through reaction with O_2.

Biochemical Carbon Concentrating Mechanisms

Biochemical CCMs concentrate carbon dioxide in one temporal or spacial region, through metabolite exchange. C4 and CAM photosynthesis both use the enzyme Phosphoenolpyruvate carboxylase (PEPC) to add CO2 to a 3-Carbon sugar. PEPC is faster than Rubisco, and more selective for CO_2.

C4

C4 plants capture carbon dioxide in their mesophyll cells (using an enzyme called Phosphoenolpyruvate carboxylase which catalyzes the combination of carbon dioxide with a compound called Phosphoenolpyruvate (PEP)), forming oxaloacetate. This oxaloacetate is then converted to malate and is transported into the bundle sheath cells (site of carbon dioxide fixation by RuBisCO) where oxygen concentration is low to avoid photorespiration. Here, carbon dioxide is removed from the malate and combined with RuBP by Rubisco in the usual way, and the Calvin cycle proceeds as normal. The CO2 concentrations in the Bundle Sheath are approximately 10-20 fold higher than the concentration in the mesophyll cells.

This ability to avoid photorespiration makes these plants more hardy than other plants in dry and hot environments, wherein stomata are closed and internal carbon dioxide levels are low. Under these conditions, photorespiration does occur in C4 plants, but at a much reduced level compared with C3 plants in the same conditions. C4 plants include sugar cane, corn (maize), and sorghum.

CAM (Crassulacean Acid Metabolism)

CAM plants, such as cacti and succulent plants, also use the enzyme PEP carboxylase to capture carbon dioxide, but only at night. Crassulacean acid metabolism allows plants to conduct most of their gas exchange in the cooler night-time air, sequestering carbon in 4 carbon sugars which can be released to the photosynthesizing cells during the day. This allows CAM plants to reduce water loss (transpiration) by maintaining closed stomata during the day. CAM plants usually display other water-saving characteristics, such as thick cuticles, stomata with small apertures, and typically lose around 1/3 of the amount of water per CO_2 fixed.

Algae

There have been some reports of algae operating a biochemical CCM: shuttling metabolites within single cells to concentrate CO_2 in one area. This process is not fully understood and may not be biologically widespread.

Biophysical Carbon Concentrating Mechanisms

This type of carbon concentrating mechanism (CCM) relies on a contained compartment within the cell into which CO_2 is shuttled, and where RuBisCO is highly expressed. In many species, biophysical CCMs are only induced under low carbon dioxide concentrations. Biophysical CCMs are more evolutionarily ancient than biochemical CCMs. There is some debate as to when biophysical CCMs first evolved, but it is likely to have been during a period of low carbon dioxide, after the great oxidation event (2.4 billion years ago). Low CO_2 periods occurred around 750, 650, and 320-270 million years ago.

Eukaryotic Algae

In nearly all species of eukaryotic algae (*Chloromonas* being one notable exception), upon induction of the CCM, ~95% of RuBisCO is densely packed into a single subcellular compartment: the pyrenoid. Carbon dioxide is concentrated in this compartment using a combination of CO_2 pumps, bicarbonate pumps, and carbonic anhydrases. The pyrenoid is not a membrane bound compartment, but is found within the chloroplast, often surrounded by a starch sheath (which is not thought to serve a function in the CCM).

Hornworts

Certain species of Hornwort are the only land plants which are known to have a biophysical CCM involving concentration of carbon dioxide within pyrenoids in their chloroplasts.

Cyanobacteria

Cyanobacterial CCMs are similar in principle to those found in Eukaryotic algae and Hornworts, but the compartment into which carbon dioxide is concentrated has several structural differences. Instead of the pyrenoid, cyanobacteria contain Carboxysomes, which have a protein shell, and linker proteins packing RuBisCO inside with a very regular structure. Cyanobacterial CCMs are much better understood than those found in Eukaryotes, partly due to the ease of genetic manipulation of prokaryotes.

Photorespiration may Serve Some Purpose

Reducing photorespiration may not result in increased growth rates for plants. Some research has suggested, for example, that photorespiration may be necessary for the assimilation of nitrate from soil. Thus, a reduction in photorespiration by genetic engineering or because of increasing atmospheric carbon dioxide due to fossil fuel burning may not benefit plants as has been proposed. Several physiological processes may be responsible for linking photorespiration and nitrogen assimilation: one is that photorespiration increases availability of NADH, which is required for the conversion of nitrate to nitrite; another link is that certain nitrite transporters also transport bicarbonate, and elevated CO_2 has been shown to suppress nitrite transport into chloroplasts.

Although photorespiration is greatly reduced in C4 species, it is still an essential pathway - mutants without functioning 2-phosphoglycolate metabolism cannot grow in normal conditions. One mutant was shown to rapidly accumulate glycolate.

Although the functions of photorespiration remain controversial, it is widely accepted that this pathway influences a wide range of processes from bioenergetics, photosystem II function, and carbon metabolism to nitrogen assimilation and respiration. The photorespiratory pathway is a major source of H2O2 in photosynthetic cells. Through H2O2 production and pyridine nucleotide interactions, photorespiration makes a key contribution to cellular redox homeostasis. In so doing, it influences multiple signalling pathways, in particular, those that govern plant hormonal responses controlling growth, environmental and defense responses, and programmed cell death.

Another theory postulates that it may function as a "safety valve", preventing the excess of reductive potential coming from an overreduced NADPH-pool from reacting with oxygen and producing free radicals, as these can damage the metabolic functions of the cell by subsequent oxidation of membrane lipids, proteins or nucleotides.

6

Hydroponics

Hydroponics is a subset of hydroculture and is a method of growing plants using mineral nutrient solutions, in water, Terrestrial plants may be grown with their roots in the mineral nutrient solution only or in an inert medium, such as perlite, gravel, biochar, mineral wool, expanded clay pebbles or coconut husk.

Researchers discovered in the 18th century that plants absorb essential mineral nutrients as inorganic ions in water. In natural conditions, soil acts as a mineral nutrient reservoir but the soil itself is not essential to plant growth. When the mineral nutrients in the soil dissolve in water, plant roots are able to absorb them. When the required mineral nutrients are introduced into a plant's water supply artificially, soil is no longer required for the plant to thrive. Almost any terrestrial plant will grow with hydroponics. Hydroponics is also a standard technique in biology research and teaching.

History

The earliest published work on growing terrestrial plants without soil was the 1627 book *Sylva Sylvarum* by Francis Bacon, printed a year after his death. Water culture became a popular research technique after that. In 1699, John Woodward published his water culture experiments with spearmint. He found that plants in less-pure water sources grew better than plants in distilled water. By 1842, a list of nine elements believed to be essential to plant growth had been compiled, and the discoveries of the German botanists Julius von Sachs and

Wilhelm Knop, in the years 1859-65, resulted in a development of the technique of soilless cultivation. Growth of terrestrial plants without soil in mineral nutrient solutions was called solution culture. It quickly became a standard research and teaching technique and is still widely used today. Solution culture is now considered a type of hydroponics where there is no inert medium.

In 1929, William Frederick Gericke of the University of California at Berkeley began publicly promoting that solution culture be used for agricultural crop production. He first termed it aquaculture but later found that aquaculture was already applied to culture of aquatic organisms. Gericke created a sensation by growing tomato vines twenty-five feet high in his back yard in mineral nutrient solutions rather than soil. By analogy with the ancient Greek term for agriculture, ãåùðïíéêÜ, the science of cultivating the earth, Gericke coined the term *hydroponics* in 1937 for the culture of plants in water.

Reports of Gericke's work and his claims that hydroponics would revolutionize plant agriculture prompted a huge number of requests for further information. Gericke had been denied use of the University's greenhouses for his experiments due to the administration's skepticism, and when the University tried to compel him to release his preliminary nutrient recipes developed at home he requested greenhouse space and time to improve them using appropriate research facilities. While he was eventually provided greenhouse space, the University assigned Hoagland and Arnon to re-develop Gericke's formula and show it held no benefit over soil grown plant yields, a view held by Hoagland. In 1940, he published the book, *Complete Guide to Soilless Gardening,* after leaving his academic position in a climate that was politically unfavourable.

Two other plant nutritionists at the University of California were asked to research Gericke's claims. Dennis R. Hoagland and Daniel I. Arnon wrote a classic 1938 agricultural bulletin, *The Water Culture Method for Growing Plants Without Soil,*. Hoagland and Arnon claimed that hydroponic crop yields were no better than crop yields with good-quality soils. Crop yields were ultimately limited by factors other than mineral nutrients, especially light. This research, however, overlooked the fact that hydroponics has other advantages including the fact that the roots of the plant have constant access to oxygen and that the plants have access to as much or as little water as they need. This is important as one of the most common errors when growing is over- and under-watering; and hydroponics prevents this from occurring as large amounts of water can be made available to the plant and any water not used, drained away, recirculated, or actively aerated, eliminating anoxic conditions, which drown root systems in soil. In soil, a grower needs to be very experienced to know exactly how much water to feed the plant. Too much and the plant will not be able to access oxygen; too little and the plant will lose the ability to transport nutrients, which are typically moved into the roots while in solution. These two researchers developed several formulas for mineral nutrient solutions, known as Hoagland solution. Modified Hoagland solutions are still used today.

One of the earliest successes of hydroponics occurred on Wake Island, a rocky atoll in the Pacific Ocean used as a refuelling stop for Pan American Airlines. Hydroponics was used there in the 1930s to grow vegetables for the passengers. Hydroponics was a necessity on Wake Island because there was no soil, and it was prohibitively expensive to airlift in fresh vegetables.

In the 1960s, Allen Cooper of England developed the Nutrient film technique. The Land Pavilion at Walt Disney World's EPCOT Centre opened in 1982 and prominently features a variety of hydroponic techniques. In recent decades, NASA has done extensive hydroponic research for their Controlled Ecological Life Support System or CELSS. Hydroponics intended to take place on Mars are using LED lighting to grow in different colour spectrum with much less heat.

Origin

Soilless Culture: Gericke originally defined hydroponics as crop growth in mineral nutrient solutions. Hydroponics is a subset of soilless culture. Many types of soilless culture do not use the mineral nutrient solutions required for hydroponics. Plants that are not traditionally grown in a climate would be possible to grow using a controlled environment system like hydroponics. NASA has also looked to utilize hydroponics in the space programme. Ray Wheeler, plant physiologist at Kennedy Space Centre's Space Life Science Lab, believes that hydroponics will create advances within space travel. He terms this as a bioregenerative life support system.

Advantages and Disadvantages

Advantages: Some of the reasons why hydroponics is being adapted around the world for food production are the following:

- No soil is needed for hydroponics.
- The water stays in the system and can be reused - thus, a lower water requirement.
- It is possible to control the nutrition levels in their entirety - thus, lower nutrition requirements.

- No nutrition pollution is released into the environment because of the controlled system.
- Stable and high yields.
- Pests and diseases are easier to get rid of than in soil because of the container's mobility.
- Ease of harvesting.
- No pesticide damage.

Today, hydroponics is an established branch of horticulture. Progress has been rapid, and results obtained in various countries have proved it to be thoroughly practical and to have very definite advantages over conventional methods of horticulture. There are two chief merits of the soil-less cultivation of plants. First, hydroponics may potentially produce much higher crop yields. Also, hydroponics can be used in places where in-ground agriculture or gardening are not possible.

Disadvantages: Without soil as a buffer, any failure to the hydroponic system leads to rapid plant death. Other disadvantages include pathogen attacks such as damp-off due to Verticillium wilt caused by the high moisture levels associated with hydroponics and over watering of soil based plants. Also, many hydroponic plants require different fertilizers and containment systems.

Techniques

The two main types of hydroponics are solution culture and medium culture. Solution culture does not use a solid medium for the roots, just the nutrient solution. The three main types of solution cultures are static solution culture, continuous-flow solution culture and aeroponics. The medium culture method has a solid medium for the roots and is named for the type of medium, e.g., sand culture, gravel culture, or rockwool culture.

There are two main variations for each medium, sub-irrigation and top irrigation. For all techniques, most hydroponic reservoirs are now built of plastic, but other materials have been used including concrete, glass, metal, vegetable solids, and wood. The containers should exclude light to prevent algae growth in the nutrient solution.

Static Solution Culture

In static solution culture, plants are grown in containers of nutrient solution, such as glass Mason jars (typically, in-home applications), plastic buckets, tubs, or tanks. The solution is usually gently aerated but may be un-aerated. If un-aerated, the solution level is kept low enough that enough roots are above the solution so they get adequate oxygen.

A hole is cut in the lid of the reservoir for each plant. There can be one to many plants per reservoir. Reservoir size can be increased as plant size increases. A home made system can be constructed from plastic food containers or glass canning jars with aeration provided by an aquarium pump, aquarium airline tubing and aquarium valves. Clear containers are covered with aluminium foil, butcher paper, black plastic, or other material to exclude light, thus helping to eliminate the formation of algae.

The nutrient solution is changed either on a schedule, such as once per week, or when the concentration drops below a certain level as determined with an electrical conductivity metre. Whenever the solution is depleted below a certain level, either water or fresh nutrient solution is added, A Mariotte's bottle, or a float valve, can be used to automatically maintain the solution level. In raft solution culture, plants are placed in a sheet of buoyant plastic that is floated on the surface of the nutrient solution. That way, the solution level never drops below the roots.

Continuous-flow Solution Culture

In continuous-flow solution culture, the nutrient solution constantly flows past the roots. It is much easier to automate than the static solution culture because sampling and adjustments to the temperature and nutrient concentrations can be made in a large storage tank that has potential to serve thousands of plants. A popular variation is the nutrient film technique or NFT, whereby a very shallow stream of water containing all the dissolved nutrients required for plant growth is recirculated past the bare roots of plants in a watertight thick root mat, which develops in the bottom of the channel and has an upper surface that, although moist, is in the air. Subsequent to this, an abundant supply of oxygen is provided to the roots of the plants.

A properly designed NFT system is based on using the right channel slope, the right flow rate, and the right channel length. The main advantage of the NFT system over other forms of hydroponics is that the plant roots are exposed to adequate supplies of water, oxygen, and nutrients. In all other forms of production, there is a conflict between the supply of these requirements, since excessive or deficient amounts of one results in an imbalance of one or both of the others. NFT, because of its design, provides a system where all three requirements for healthy plant growth can be met at the same time, provided that the simple concept of NFT is always remembered and practised. The result of these advantages is that higher yields of high-quality produce are obtained over an extended period of cropping. A downside of NFT is that it has very little buffering against interruptions in the flow, e.g., power outages. But, overall, it is probably one of the more productive techniques.

The same design characteristics apply to all conventional NFT systems. While slopes along channels of 1:100 have been recommended, in practice it is difficult to build a base for channels that is sufficiently true to enable nutrient films to flow without ponding in locally depressed areas. As a consequence, it is recommended that slopes of 1:30 to 1:40 are used. This allows for minor irregularities in the surface, but, even with these slopes, ponding and water logging may occur. The slope may be provided by the floor, or benches or racks may hold the channels and provide the required slope. Both methods are used and depend on local requirements, often determined by the site and crop requirements.

As a general guide, flow rates for each gully should be 1 litre per minute. At planting, rates may be half this and the upper limit of 2 L/min appears about the maximum. Flow rates beyond these extremes are often associated with nutritional problems. Depressed growth rates of many crops have been observed when channels exceed 12 metres in length. On rapidly growing crops, tests have indicated that, while oxygen levels remain adequate, nitrogen may be depleted over the length of the gully. As a consequence, channel length should not exceed 10–15 metres. In situations where this is not possible, the reductions in growth can be eliminated by placing another nutrient feed halfway along the gully and reducing flow rates to 1 L/min through each outlet.

Aeroponics

Aeroponics is a system wherein roots are continuously or discontinuously kept in an environment saturated with fine drops (a mist or aerosol) of nutrient solution. The method requires no substrate and entails growing plants with their roots suspended in a deep air or growth chamber with the roots periodically wetted with a fine mist of atomized nutrients. Excellent aeration is the main advantage of aeroponics. Aeroponic techniques have proven to be commercially successful for propagation, seed germination, seed potato production, tomato production, leaf crops, and micro-greens. Since inventor Richard Stoner commercialized aeroponic technology in 1983, aeroponics has been implemented as an alternative to water intensive hydroponic systems worldwide. The limitation of hydroponics is the fact that 1 kg of water can only hold 8 mg of air, no matter whether aerators are utilized or not.

Another distinct advantage of aeroponics over hydroponics is that any species of plants can be grown in a true aeroponic system because the micro environment of an aeroponic can be finely controlled. The limitation of hydroponics is that only certain species of plants can survive for so long in water before they become waterlogged. The advantage of aeroponics is that suspended aeroponic plants receive 100% of the available oxygen and carbon dioxide to the roots zone, stems, and leaves, thus accelerating biomass growth and reducing rooting times. NASA research has shown that aeroponically grown plants have an 80% increase in dry weight biomass (essential minerals) compared to hydroponically grown plants. Aeroponics used 65% less water than hydroponics. NASA also concluded that aeroponically grown plants requires ¼ the nutrient input compared to hydroponics. Unlike hydroponically grown plants, aeroponically grown plants will not suffer transplant shock when transplanted to soil, and offers growers the ability to reduce the spread of disease and pathogens. Aeroponics is also widely used in laboratory studies of plant physiology and plant pathology. Aeroponic techniques have been given special attention from NASA since a mist is easier to handle than a liquid in a zero gravity environment.

Passive Sub-irrigation

Passive sub-irrigation, also known as passive hydroponics or semi-hydroponics, is a method wherein plants are grown in an inert porous medium that transports water and fertilizer to the roots by capillary action from a separate reservoir as necessary, reducing labour and providing a constant supply of water to the roots. In the simplest method, the pot sits in a shallow solution of fertilizer and water or on a capillary mat saturated with nutrient solution. The various hydroponic media available, such as expanded clay and coconut husk, contain more air space than more traditional potting mixes, delivering increased oxygen to the roots, which is important in epiphytic plants such as orchids and bromeliads, whose roots are exposed to the air in nature. Additional advantages of passive hydroponics are the reduction of root rot and the additional ambient humidity provided through evaporations.

Ebb and Flow or Flood and Drain Sub-irrigation

In its simplest form, there is a tray above a reservoir of nutrient solution. Either the tray is filled with growing medium (clay granules being the most common) and planted directly or pots of medium stand in the tray. At regular intervals, a simple timer causes a pump to fill the upper tray with nutrient solution, after which the solution drains back down into the reservoir. This keeps the medium regularly flushed with nutrients and air. Once the upper tray fills past the drain stop, it begins recirculating the water until the timer turns the pump off, and the water in the upper tray drains back into the reservoirs.

Run to Waste

In a run to waste system, nutrient and water solution is periodically applied to the medium surface. This may be done in its simplest form, by manually applying a nutrient-and-water solution one or more times per day in a container of inert growing media, such as rockwool, perlite, vermiculite, coco fibre, or sand. In a slightly more complex system, it is automated with a delivery pump, a timer and irrigation tubing to deliver nutrient solution with a delivery frequency that is governed by the key parametres of plant size, plant growing stage, climate, substrate, and substrate conductivity, pH, and water content.

In a commercial setting, watering frequency is multi factorial and governed by computers or PLCs. Commercial hydroponics production of large plants like tomatoes, cucumber, and peppers use one form or another of run to waste hydroponics. In environmentally responsible uses, the nutrient rich waste is collected and processed through an on site filtration system to be used many times, making the system very productive.

Deep Water Culture

The hydroponic method of plant production by means of suspending the plant roots in a solution of nutrient-rich, oxygenated water. Traditional methods favour the use of plastic buckets and large containers with the plant contained in a net pot suspended from the centre of the lid and the roots suspended in the nutrient solution. The solution is oxygen saturated from an air pump combined with porous stones. With this method, the plants grow much faster because of the high amount of oxygen that the roots receive.

Bubbleponics

"Bubbleponics" is the art of delivering highly oxygenated nutrient solution direct to the root zone of plants. While Deep Water Culture involves the plant roots hanging down into a reservoir of water below, the term Bubbleponics describes a top-fed Deep Water Culture (DWC) hydroponic system. In this method, the water is pumped from the reservoir up to the roots (top feeding). The water is released over the plant's roots and then runs back into the reservoir below in a constantly recirculating system. As with Deep Water Culture, there is an airstone in the reservoir that pumps air into the water via a hose from outside the reservoir. The airstone helps add oxygen to the water. Both the airstone and the water pump run 24 hours a day.

The biggest advantages with Bubbleponics over Deep Water Culture involve increased growth during the first few weeks. With Deep Water

Culture, there is a time where the roots have not reached the water yet. With Bubbleponics, the roots get easy access to water from the beginning and will grow to the reservoir below much more quickly than with a Deep Water Culture system. Once the roots have reached the reservoir below, there is not a huge advantage with Bubbleponics over Deep Water Culture. However, due to the quicker growth in the beginning, a few weeks of grow time can be shaved off.

Fogponics

Fogponics is an advanced form of aeroponics which uses water in a vaporised form to transfer nutrients and oxygen to enclosed suspended plant roots. Using the same general idea behind aeroponics except fogponics utilize a 5-30 μm mist within the rooting chamber and as use for a foliar feeding mechanism. Plants best absorb particles from the 1-25 μm range, the smaller particulate size means faster absorption. The added benefit of using fogponic's over traditional hydroponics systems is that the plants require less energy in root growth and mass, and are able to still sustain a large plant.

Trials using fogponics, have shown that water and energy use is reduced compared with traditional aeroponic and hydroponic systems, and that higher yields are achievable.

Rotary

A rotary hydroponic garden is a style of commercial hydroponics created within a circular frame which rotates continuously during the entire growth cycle of whatever plant is being grown. While system specifics vary, systems typically rotate once per hour, giving a plant 24 full turns within the circle each 24 hour period. Within the centre of each rotary hydroponic garden is a high intensity grow light, designed to simulate sunlight, often with the assistance of a mechanized timer.

Each day, as the plants rotate, they are periodically watered with a hydroponic growth solution to provide all nutrient necessary for robust growth. Due to the plants continuous fight against gravity plants typically mature much more quickly than when grown in soil or other traditional hydroponic growing systems. Due to the small foot print a rotary hydroponic system has, it allows for more plant material to be grown per sq foot of floor space than other traditional hydroponic systems.

Substrates

One of the most obvious decisions hydroponic farmers have to make is which medium they should use. Different media are appropriate for different growing techniques.

Expanded Clay Aggregate

***Figure:** Expanded clay pebbles.*

Baked clay pellets, are suitable for hydroponic systems in which all nutrients are carefully controlled in water solution. The clay pellets are inert, pH neutral and do not contain any nutrient value.

The clay is formed into round pellets and fired in rotary kilns at 1,200 °C (2,190 °F). This causes the clay to expand, like popcorn, and become porous. It is light in weight, and does not compact over time. The shape of an individual pellet can be irregular or uniform depending on brand and manufacturing process. The manufacturers consider expanded clay to be an ecologically sustainable and re-usable growing medium because of its ability to be cleaned and sterilized, typically by washing in solutions of white vinegar, chlorine bleach, or hydrogen peroxide ($H2O2$), and rinsing completely. Another view is that clay pebbles are best not re-used even when they are cleaned, due to root growth that may enter the medium. Breaking open a clay pebble after a crop has been grown will reveal this growth.

Growstones

Growstones, made from glass waste, have both more air and water retention space than perlite and peat. This aggregate holds more water than parboiled rice hulls.

Coir

Coco Peat, also known as coir or coco, is the leftover material after the fibres have been removed from the outermost shell (bolster) of the coconut. Coir is a 100% natural grow and flowering medium. Coconut

Coir is colonized with trichoderma Fungi, which protects roots and stimulates root growth. It is extremely difficult to over water coir due to its perfect air-to-water ratio, plant roots thrive in this environment, coir has a high cation exchange, meaning it can store unused minerals to be released to the plant as and when it requires it. Coir is available in many forms, most common is coco peat, which has the appearance and texture of soil but contains no mineral content.

Rice Hulls

Parboiled rice hulls (PBH) decay over time. Rice hulls allow drainage, and even retain less water than growstones. A study showed that rice hulls didn't affect the effects of plant growth regulators. Rice hulls are an agricultural byproduct that would otherwise have little use.

Perlite

Perlite is a volcanic rock that has been superheated into very lightweight expanded glass pebbles. It is used loose or in plastic sleeves immersed in the water. It is also used in potting soil mixes to decrease soil density. Perlite has similar properties and uses to vermiculite but, in general, holds more air and less water. If not contained, it can float if flood and drain feeding is used. It is a fusion of granite, obsidian, pumice and basalt. This volcanic rock is naturally fused at high temperatures undergoing what is called "Fusionic Metamorphosis".

Pumice

Like perlite, pumice is a lightweight, mined volcanic rock that finds application in hydroponics.

Vermiculite

Like perlite, vermiculite is a mineral that has been superheated until it has expanded into light pebbles. Vermiculite holds more water than perlite and has a natural "wicking" property that can draw water and nutrients in a passive hydroponic system. If too much water and not enough air surrounds the plants roots, it is possible to gradually lower the medium's water-retention capability by mixing in increasing quantities of perlite.

Sand

Sand is cheap and easily available. However, it is heavy, does not hold water very well, and it must be sterilized between use.

Gravel

The same type that is used in aquariums, though any small gravel can be used, provided it is washed first. Indeed, plants growing in a typical traditional gravel filter bed, with water circulated using electric powerhead pumps, are in effect being grown using gravel hydroponics. Gravel is inexpensive, easy to keep clean, drains well and will not become waterlogged. However, it is also heavy, and, if the system does not provide continuous water, the plant roots may dry out.

Wood Fibre

Wood fibre, produced from steam friction of wood, is a very efficient organic substrate for hydroponics. It has the advantage that it keeps its structure for a very long time. Wood fibre has been shown to reduce the effects of "plant growth regulators."

Sheep Wool

Wool from shearing sheep is a little-used yet promising renewable growing medium. In a study comparing wool with peat slabs, coconut fibre slabs, perlite and rockwool slabs to grow cucumber plants, sheep wool had a greater air capacity of 70%, which decreased with use to a comparable 43%, and water capacity that increased from 23% to 44% with use. Using sheep wool resulted in the greatest yield out of the tested substrates, while application of a biostimulator consisting of humic acid, lactic acid and Bacillus subtilis improved yields in all substrates.

Rock Wool

Rock wool (mineral wool) is the most widely used medium in hydroponics. Rock wool is an inert substrate suitable for both run to waste and recirculating systems. Rock wool is made from molten rock, basalt or 'slag' that is spun into bundles of single filament fibres, and bonded into a medium capable of capillary action, and is, in effect, protected from most common microbiological degradation. Rock wool has many advantages and some disadvantages. The latter being the possible skin irritancy (mechanical) whilst handling (1:1000). Flushing with cold water usually brings relief.

Advantages include its proven efficiency and effectiveness as a commercial hydroponic substrate. Most of the rock wool sold to date is a non-hazardous, non-carcinogenic material, falling under Note Q of the European Union Classification Packaging and Labelling Regulation (CLP).

Brick Shards

Brick shards have similar properties to gravel. They have the added disadvantages of possibly altering the pH and requiring extra cleaning before reuse.

Polystyrene Packing Peanuts

Polystyrene packing peanuts are inexpensive, readily available, and have excellent drainage. However, they can be too lightweight for some uses. They are used mainly in closed-tube systems. Note that polystyrene peanuts must be used; biodegradable packing peanuts will decompose into a sludge. Plants may absorb styrene and pass it to their consumers; this is a possible health risk.

Nutrient Solutions

Plant nutrients used in hydroponics are dissolved in the water and are mostly in inorganic and ionic form. Primary among the dissolved cations (positively charged ions) are Ca^{2+} (calcium), Mg2+(magnesium), and K+(potassium); the major nutrient anions in nutrient solutions are NO–3 (nitrate), SO2–4 (sulfate), and H2PO–4 (dihydrogen phosphate).

Numerous 'recipes' for hydroponic solutions are available. Many use different combinations of chemicals to reach similar total final compositions. Commonly used chemicals for the macronutrients include potassium nitrate, calcium nitrate, potassium phosphate, and magnesium sulfate. Various micronutrients are typically added to hydroponic solutions to supply essential elements; among them are Fe (iron), Mn (manganese), Cu (copper), Zn (zinc), B (boron), Cl (chlorine), and Ni (nickel). Chelating agents are sometimes used to keep Fe soluble, and humic acids can be added to increase nutrient uptake. Many variations of the nutrient solutions used by Arnon and Hoagland have been styled 'modified Hoagland solutions' and are widely used. Variation of different mixes throughout the plant life-cycle, further optimizes its nutritional value. Plants will change the composition of the nutrient solutions upon contact by depleting specific nutrients more rapidly than others, removing water from the solution, and altering the pH by excretion of either acidity or alkalinity. Care is required not to allow salt concentrations to become too high, nutrients to become too depleted, or pH to wander far from the desired value.

Although pre-mixed concentrated nutrient solutions are generally purchased from commercial nutrient manufacturers by hydroponic hobbyists and small commercial growers, several tools exists to help anyone prepare their own solutions without extensive knowledge about chemistry. The free and open source tools HydroBuddy and HydroCal have been created by professional chemists to help any hydroponics grower prepare their own nutrient solutions. The first programme is available for Windows, Mac and Linux while the second one can be used through a simple Java interface. Both programmes allow for basic nutrient solution preparation although HydroBuddy provides added functionality to use and save custom substances, save formulations and predict electrical conductivity values.

The well-oxygenated and enlightened environment promotes the development of algae. It is therefore necessary to wrap the tank with black film obscuring all light.

Organic hydroponics uses the solution containing microorganisms. In organic hydroponics, organic fertilizer can be added in the hydroponic solution because microorganisms degrade organic fertilizer into inorganic nutrients. In contrast, conventional hydroponics cannot use organic fertilizer because organic compounds in the hydroponic solution show phytotoxic effects.

Commercial

Some commercial installations use no pesticides or herbicides, preferring integrated pest management techniques. There is often a price premium willingly paid by consumers for produce that is labelled "organic". Some states in the USA require soil as an essential to obtain organic certification. There are also overlapping and somewhat contradictory rules established by the US Federal Government, so some food grown with hydroponics can be certified organic. Most hydroponically grown produce cannot be sold as organic due to the fact that they do not use soil as a growing medium.

Hydroponics also saves water; it uses as little as ^{1}D $_{20}$ the amount as a regular farm to produce the same amount of food. The water table can be impacted by the water use and run-off of chemicals from farms, but hydroponics may minimize impact as well as having the advantage that water use and water returns are easier to measure. This can save the farmer money by allowing reduced water use and the ability to measure consequences to the land around a farm.

To increase plant growth, lighting systems such as metal-halide lamp for growing stage only or high-pressure sodium for growing/ flowering/blooming stage are used to lengthen the day or to supplement natural sunshine if it is scarce. Metal halide emits more light in the blue spectrum, making it ideal for plant growth but is harmful to unprotected skin and can cause skin cancer. High-pressure sodium emits more light in the red spectrum, meaning that it is best suited for supplementing natural sunshine and can be used throughout the growing cycle. However, these lighting systems require large amounts of electricity to operate, making efficiency and safety very critical.

The environment in a hydroponics greenhouse is tightly controlled for maximum efficiency, and this new mindset is called soil-less/ controlled-environment agriculture (CEA). With this growers can make ultra-premium foods anywhere in the world, regardless of temperature and growing seasons. Growers monitor the temperature, humidity, and pH level constantly.

Hydroponics have been used to enhance vegetables to provide more nutritional value. A hydroponic farmer in Virginia has developed a calcium and potassium enriched head of lettuce, scheduled to be widely available in April 2007. Grocers in test markets have said that the lettuce sells "very well", and the farmers claim that their hydroponic lettuce uses 90% less water than traditional soil farming.

Advancements

With pest problems reduced, and nutrients constantly fed to the roots, productivity in hydroponics is high, although plant growth can be limited by the low levels of carbon dioxide in the atmosphere, or limited light exposure. To increase yield further, some sealed greenhouses inject carbon dioxide into their environment to help growth (CO_2 enrichment), add lights to lengthen the day, or control vegetative growth, etc.

Vertical Farming

Vertical farming is cultivating plant or animal life within a skyscraper greenhouse or on vertically inclined surfaces. The modern idea of vertical farming uses techniques similar to glass houses, where natural sunlight can be augmented with artificial lighting.

Types

"Vertical farming" was coined by Gilbert Ellis Bailey in 1915 in his book *Vertical Farming*. This was not the current meaning—he wrote about farming underground with the use of explosives. Modern usage refers to skyscrapers using some degree of natural light.

Mixed-use Skyscrapers

Mixed-use skyscrapers were proposed and built by architect Ken Yeang. Yeang proposes that instead of hermetically sealed mass-produced agriculture that plant life should be cultivated within open air, mixed-use skyscrapers for climate control and consumption (i.e. a personal or communal planting space as per the needs of the individual). This version of vertical farming is based upon personal or community use rather than the wholesale production and distribution plant and animal life that aspires to feed an entire city. It thus requires less of an initial investment than Despommier's "The Vertical Farm". However, neither Despommier nor Yeang are the conceptual "originators", nor is Yeang the inventor of vertical farming in skyscrapers.

Despommier's Skyscrapers

Ecologist Dickson Despommier argues that vertical farming is legitimate for environmental reasons. He claims that the cultivation of plant and animal life within skyscrapers will produce less embedded energy and toxicity than plant and animal life produced on natural landscapes. He moreover claims that natural landscapes are too toxic for natural, agricultural production, despite the ecological and environmental costs of extracting materials to build skyscrapers for the simple purpose of agricultural production.

Vertical farming according to Despommier thus discounts the value of natural landscape in exchange for the idea of "skyscraper as spaceship". Plant and animal life are mass-produced within hermetically sealed, artificial environments that have little to do with the outside world. In this sense, they could be built anywhere regardless of the context. This is unlikely to be advantageous with regards to energy consumption as the internal environment must be maintained to sustain life within the skyscraper. However, this is not necessarily the case, as one of the most important features of a vertical farm is that it would contain some form of renewable energy technology, be it solar panels, wind turbines, or a water capture system, and could contain all three. The vertical farm is designed to be sustainable, and to enable nearby inhabitants to work at the farm. Despommier's concept of "The Vertical Farm" emerged in 1999 at Columbia University. It promotes the mass cultivation of plant and animal life for commercial purposes in skyscrapers.

History

A commercial high-rise farm such as 'The Vertical Farm' has never been built, yet extensive photographic documentation and several historical books on the subject suggest that research on the subject was not diligently pursued. New sources indicate that a tower hydroponicum existed in Armenia prior to 1951.

Proponents argue that, by allowing traditional outdoor farms to revert to a natural state and reducing the energy costs needed to transport foods to consumers, vertical farms could significantly alleviate climate change produced by excess atmospheric carbon. Critics have noted that the costs of the additional energy needed for artificial lighting, heating and other vertical farming operations would outweigh the benefit of the building's close proximity to the areas of consumption.

One of the earliest drawings of a tall building that cultivates food for the purposes of consumption was published as early as Life Magazine 1909. The reproduced drawings feature vertically stacked homesteads set amidst a farming landscape. This proposal can be seen in Rem Koolhaas's Delirious New York. Koolhaas wrote that this 1909 theorem is 'The Skyscraper as Utopian device for the production of unlimited numbers of virgin sites on a metropolitan location' (1994, 82).

Other architectural proposals that provide the seeds for the Vertical Farm project include Le Corbusier's Immeubles-Villas (1922) and SITE's Highrise of homes (1972). SITE's Highrise of homes, is a near revival of the 1909 Life Magazine Theorem. In fact, built examples of tower hydroponicums are quite well documented in the canonical text of "The Glass House" by John Hix. Images of the vertical farms at the School of Gardeners in Langenlois, Austria, and the glass tower at the Vienna International Horticulture Exhibition (1964) clearly show that vertical farms existed more than 40 years prior to contemporary discourse on the subject.

Although architectural precedents remain valuable, the technological precedents that make vertical farming possible can be traced back to horticultural history through the development of greenhouse and hydroponic technology. Early building types or Hydroponicums were developed, integrating hydroponic technology into building systems. These horticultural building systems evolved from greenhouse technology, and paved the way for the modern concept of the vertical farm. The British Interplanetary Society developed a hydroponicum for lunar conditions and other building prototypes were developed during the early days of space exploration. During this era of expansion and experimentation, the first Tower Hydroponic Units were developed in Armenia.

The Armenian tower hydroponicums are the first built examples of a vertical farm, and is documented in Sholto Douglas' seminal text "Hydroponics: The Bengal System" first published in 1951 with data from the then-East Pakistan, today's Bangladesh, and the Indian state of West Bengal. Contemporary notions of vertical farming are predated by this early technology by more than 50 years. Contemporary precursors that have been published, or built, are Ken Yeang's Bioclimatic Skyscraper (Menara Mesiniaga, built 1992); MVRDV's PigCity, 2000; MVRDV's Meta City/ Datatown (1998–2000); Pich-Aguilera's Garden Towers (2001).

Ken Yeang is perhaps the most widely known architects that has promoted the idea of the 'mixed-use' Bioclimatic Skyscraper which combines living units and opportunities for food production.

Early prototypes of vertical farms, or "Tower Hydroponicums" existed in Armenia prior to 1951 during an era of hydroponic and horticultural building system research fueled by space exploration and a transatlantic technology race.

The latest version of these very idea is Dickson Despommier's "The Vertical Farm".

Dickson Despommier, a professor of environmental health sciences and microbiology at Columbia University in New York City, modernized the idea of vertical farming in 1999 with graduate students in a medical ecology class. Although much of Despommier's suggestions have been challenged and strongly criticized from an environmental science and engineering point of view, the idea's popularization in recent years has been largely the result of Despommier's assertion that food production can be transformed.

Despommier had originally challenged his class to feed the population of Manhattan (About 2,000,000 people) using 13 acres (5.3 ha) of usable rooftop gardens. The class calculated that, by using rooftop gardening methods, only 2 percent would be fed. Unsatisfied with the results, Despommier made an off-the-cuff suggestion of growing plants indoors, vertically. The idea sparked the students' interests and gained major momentum. By 2001 the first outline of a vertical farm was introduced and today scientists, architects, and investors worldwide are working together to make the concept of vertical farming a reality. In an interview with Miller-McCune.com, Despommier described how vertical farms would function:

Each floor will have its own watering and nutrient monitoring systems. There will be sensors for every single plant that tracks how much and what kinds of nutrients the plant has absorbed. You'll even have systems to monitor plant diseases by employing DNA chip technologies that detect the presence of plant pathogens by simply sampling the air and using snippets from various viral and bacterial infections. It's very easy to do.

Moreover, a gas chromatograph will tell us when to pick the plant by analyzing which flavenoids the produce contains. These flavonoids are what gives the food the flavors you're so fond of, particularly for more aromatic produce like tomatoes and peppers. These are all right-off-the-shelf technologies. The ability to construct a vertical farm exists now. We don't have to make anything new.

Architectural designs have been produced by Chris Jacobs and Andrew Kranis from Columbia University and Gordon Graff from the

University of Waterloo's School of Architecture in Cambridge, ON. Together with Graff, and after disagreeing with Despommier's technical assumptions regarding energy and water balances in 2011, Tahbit Chowdhury and a multidisciplinary team from Waterloo's Dept. of Environmental Engineering and Dept. of Systems Design Engineering augmented the concepts with a focus on low-energy economically-intensive protein-production. Along with Chowdhury, others to have disagreed with Despommier's approach include Pierre Desrochers of the University of Toronto and Dennis T. Avery of the Centre for Global Food Issues, affiliated with the Hudson Institute.

Chowdhury and Graff applied advanced industrial engineering design philosophies to modernize current greenhouse technology as it pertains to hydroponics and aeroponics. The results of the Waterloo team's work showed that there is sufficient technical grounds to begin implementing Despommier's ideas for skyscrapers. However, Chowdhury and Graff showed that the designs will be dramatically different from what Despommier envisioned at Columbia.

Mass media attention began with an article written in *New York* magazine. Since 2007, articles have appeared in *The New York Times*, *U.S. News & World Report*, *Popular Science*, *Scientific American* and *Maxim*, among others, as well as radio and television features.

As of 2012, Vertical Harvest is working on raising funds for an urban, small-scale vertical farm in Jackson Hole, Wyoming.

Advantages

Several potential advantages of vertical farming have been discussed by Despommier. Many of these benefits are obtained from scaling up hydroponic or aeroponic growing methods.

Preparation for the Future

It is estimated that by the year 2050, close to 80% of the world's population will live in urban areas and the total population of the world will increase by 3 billion people. A very large amount of land may be required depending on the change in yield per hectare. Scientists are concerned that this large amount of required farmland will not be available and that severe damage to the earth will be caused by the added farmland. Vertical farms, if designed properly, may eliminate the need to create additional farmland and help create a cleaner environment.

Increased Crop Production

Unlike traditional farming in non-tropical areas, indoor farming can produce crops year-round. All-season farming multiplies the productivity of the farmed surface by a factor of 4 to 6 depending on the crop. With some crops, such as strawberries, the factor may be as high as 30.

Furthermore, as the crops would be sold in the same infrastructures in which they are grown, they will not need to be transported between production and sale, resulting in less spoilage, infestation, and energy required than conventional farming encounters. Research has shown that 30% of harvested crops are wasted due to spoilage and infestation, though this number is much lower in developed nations.

Despommier suggests that, if dwarf versions of certain crops are used (e.g. dwarf wheat developed by NASA, which is smaller in size but richer in nutrients), year-round crops, and "stacker" plant holders are accounted for, a 30-story building with a base of a building block (5 acres (20,000 m^2)) would yield a yearly crop analogous to that of 2,400 acres (9,700,000 m^2) of traditional farming.

Protection from Weather-related Problems

Crops grown in traditional outdoor farming suffer from the often suboptimal, and sometimes extreme, nature of geological and meteorological events such as undesirable temperatures or rainfall amounts, monsoons, hailstorms, tornadoes, flooding, wildfires, and severe droughts. The protection of crops from weather is increasingly important as global climate change occurs. "Three recent floods (in 1993, 2007 and 2008) cost the United States billions of dollars in lost crops, with even more devastating losses in topsoil. Changes in rain patterns and temperature could diminish India's agricultural output by 30 percent by the end of the century."

Because vertical farming provides a controlled environment, the productivity of vertical farms would be mostly independent of weather and protected from extreme weather events. Although the controlled environment of vertical farming negates most of these factors, earthquakes and tornadoes still pose threats to the proposed infrastructure, although this again depends on the location of the vertical farms.

Conservation of Resources

Each unit of area in a vertical farm could allow up to 20 units of area of outdoor farmland to return to its natural state, and recover farmlands due to development from original flat farmlands.

Vertical farming would reduce the need for new farmland due to overpopulation, thus saving many natural resources, currently threatened by deforestation or pollution. Deforestation and desertification caused by agricultural encroachment on natural biomes would be avoided. Because vertical farming lets crops be grown closer to consumers, it would substantially reduce the amount of fossil fuels currently used to transport and refrigerate farm produce. Producing food indoors reduces or eliminates conventional plowing, planting, and harvesting by farm machinery, also powered by fossil fuels. Burning less fossil fuel would reduce air pollution and the carbon dioxide emissions that cause climate change, as well as create healthier environments for humans and animals alike.

Organic Crops

The controlled growing environment reduces the need for pesticides, namely herbicides and fungicides. Advocates claim that producing organic crops in vertical farms is practical and the most likely production

Halting Mass Extinction

Withdrawing human activity from large areas of the Earth's land surface may be necessary to slow and eventually halt the current anthropogenic mass extinction of land animals.

Traditional agriculture is highly disruptive to wild animal populations that live in and around farmland and some argue it becomes unethical when there is a viable alternative. One study showed that wood mouse populations dropped from 25 per hectare to 5 per hectare after harvest, estimating 10 animals killed per hectare each year with conventional farming. In comparison, vertical farming would cause very little harm to wildlife.

Impact on Human Health

Traditional farming is a hazardous occupation with particular risks that often take their toll on the health of human laborers. Such risks include: exposure to infectious diseases such as malaria and schistosomes, exposure to toxic chemicals commonly used as pesticides and fungicides, confrontations with dangerous wildlife such as venomous snakes, and the severe injuries that can occur when using large industrial farming equipment.

Whereas the traditional farming environment inevitably contains these risks (particularly in the farming practice known as "slash and burn"), vertical farming – because the environment is strictly controlled and predictable – reduces some of these dangers. Currently, the American food system makes fast, unhealthy food cheap while fresh produce is less available and more expensive, encouraging poor eating habits. These poor eating habits lead to health problems such as obesity, heart disease, and diabetes.

Urban Growth

Vertical farming, used in conjunction with other technologies and socioeconomic practices, could allow cities to expand while remaining largely self-sufficient food wise. This would allow for large urban centres that could grow without destroying considerably larger areas of forest to provide food for their people. Moreover, the industry of vertical farming will provide employment to these expanding urban centres. This may help displace the unemployment created by the dismantling of traditional farms, as more farm laborers move to cities in search of work. It is highly unlikely that traditional farms will become obsolete, as there are many crops that are not suited for vertical farming, and the production costs are currently much lower.

Energy Production

Vertical farms could exploit methane digesters to generate a small portion of its own electrical needs. Methane digesters could be built on site to transform the organic waste generated at the farm into biogas which is generally composed of 65% methane along with other gases. This biogas could then be burned to generate electricity for the greenhouse.

Technologies and Devices

Vertical farming relies on the use of various physical methods to become effective. Combining these technologies and devices in an integrated whole is necessary to make Vertical Farming a reality. Various methods are proposed and under research. The most common technologies suggested are:

- Greenhouse
- The Folkewall and other vertical growing architectures
- Aeroponics / Hydroponics / Aquaponics
- Composting
- Grow light
- Phytoremediation
- Skyscraper
- Controlled-environment agriculture

Plans

Despommier argues that the technology to construct vertical farms currently exists. He also states that the system can be profitable and effective, a claim evidenced by some preliminary research posted on the project's website. Developers and local governments in the following cities have expressed serious interest in establishing a vertical farm: Incheon (South Korea), Abu Dhabi (United Arab Emirates), and Dongtan (China), New York City, Portland, Ore., Los Angeles, Las Vegas, Seattle, Surrey, B.C., Toronto, Paris, Bangalore, Dubai, Shanghai and Beijing. The Illinois Institute of Technology is now crafting a detailed plan for Chicago. It is suggested that prototype versions of vertical farms should be created first, possibly at large universities interested in the research of vertical farms, in order to prevent failures such as the Biosphere 2 project in Oracle, Arizona.

In 2009, the world's first pilot production system was installed at Paignton Zoo Environmental Park in the United Kingdom. The project showcased a technological solution for vertical farming and provided a physical base to conduct research into sustainable urban food production. The produce is used to feed the zoos animals while the project enables evaluation of the systems and provides an educational resource to advocate for change in unsustainable land use practices that impact upon global biodiversity and ecosystem services,

In 2010, the Green Zionist Alliance proposed a resolution at the 36th World Zionist Congress calling on Keren Kayemet L'Yisrael (Jewish National Fund in Israel) to develop vertical farms in Israel.

In 2012, the world's first commercial vertical farm was opened in Singapore, developed by Sky Greens Farms, and is three stories high. They currently have over 100 towers that stand at nine metres tall.

Problems

Economics: Opponents question the potential profitability of vertical farming. A detailed cost analysis of start-up costs, operation costs, and revenue has not been done. The extra cost of lighting, heating, and powering the vertical farm may negate any of the cost benefits received by the decrease in transportation expenses. The economic and environmental benefits of vertical farming rest partly on the concept of minimizing food miles, the distance that food travels from farm to consumer. However, a recent analysis suggests that transportation is only a minor contributor to the economic and environmental costs of supplying food to urban populations. The author of the report, University of Toronto professor Pierre Desrochers, concluded that "food miles are, at best, a marketing fad."

Similarly, if the power needs of the vertical farm are met by fossil fuels, the environmental effect may be a net loss; even building low-carbon capacity to power the farms may not make as much sense as simply leaving the traditional farms in place, and burning less coal.

The initial building costs will be easily over \$100 million, for a 60 hectare vertical farm. Office occupancy costs can be very high in major cities, with cities such as Tokyo, Moscow, Mumbai, Dubai, Milan, Zurich, and Sao Paulo ranging from \$1850 to \$880 per square metre, respectively.

Energy Use

During the growing season, the sun shines on a vertical surface at an extreme angle such that much less light is available to crops than when they are planted on flat land. Therefore, supplemental light, would be required in order to obtain economically viable yields. Bruce Bugbee, a crop physiologist at Utah State University, believes that the power demands of vertical farming will be too expensive and uncompetitive with traditional farms using only free natural light. The environmental writer George Monbiot calculated that the cost of providing enough supplementary light to grow the grain for a single loaf would be about \$15. An article in the Economist argued that "even though crops growing in a glass skyscraper will get some natural sunlight during the day, it won't be enough" and "the cost of powering artificial lights will make indoor farming prohibitively expensive".

As "The Vertical Farm" proposes a controlled environment, heating and cooling costs will be at least as costly as any other tower. But there also remains the issue of complicated, if not more expensive, plumbing and elevator systems to distribute food and water throughout. Even throughout the northern continental United States, while heating with relatively cheap fossil fuels, the heating cost can be over \$200,000/hectare.

To address this problem, The Plant in Chicago is building an anaerobic digester into the building. This will allow the farm to operate off the energy grid. Moreover, the anaerobic digester will be recycling waste from nearby businesses that would otherwise go into landfills.

Pollution

Regular greenhouse produce is known to create more greenhouse gases than field produce, largely due to higher energy use per kilogram of produce. With vertical farms requiring much greater energy per kilogram of produce, mainly through increased lighting, than regular greenhouses, the amount of pollution created will be much higher than that from field produce.

As plants acquire nearly all their carbon from the atmosphere, greenhouse growers commonly supplement CO_2 levels to 3-4 times the rate normally found in the atmosphere. This increase in CO_2, which has been shown to increase photosynthesis rates by 50%, contributes to the higher yields expected in vertical farming. It is not uncommon to find greenhouses burning fossil fuels purely for this purpose, as other CO_2 sources, like from furnaces, contain pollutants such as sulphur dioxide and ethylene which significantly damage plants. This means a vertical farm will require a CO_2 source, most likely from combustion, even if the rest of the farm is powered by 'green' energy. Also, through necessary ventilation, much CO_2 will be leaked into the city's atmosphere.

Greenhouse growers commonly exploit photoperiodism in plants to control whether the plants are in a vegetative or reproductive stage. As part of this control, growers will have the lights on past sunset and before sunrise or periodically throughout the night. Single story greenhouses are already a nuisance to neighbours because of light pollution, a 30 story vertical farm in a densely populated area will surely face problems because of its light pollution.

Hydroponics greenhouses regularly change the water, meaning there is a large quantity of water containing fertilizers and pesticides that must be disposed of. While solutions are currently being worked on, the most common method of simply spreading the mixture over a sufficient area of neighbouring farmland or wetlands would be more difficult for an urban vertical farm.

Terrace Garden

In gardening, a terrace is an element where a raised flat paved or gravelled section overlooks a prospect. A raised terrace keeps a house dry and provides a transition between the hard materials of the architecture and softer ones of the garden.

History

Persia: Since a level site is generally regarded as a requisite for comfort and repose, the terrace as a raised viewing platform made an early appearance in the ancient Persian gardening tradition, where the enclosed orchard, or paradise, was to be viewed from a ceremonial tent. Such a terrace had its origins in the far older agricultural practice of terracing a sloping site. The Hanging Gardens of Babylon must have been built on an artificial mountain with stepped terraces, like those on a ziggurat.

Ancient Rome

Lucullus brought back to Rome first-hand experience of Persian gardening in the hilly sites of Asia Minor; the villa gardens of Maecenas, which included libraries open to scholars, incurred the disdain of Seneca.

At Praeneste during the early Imperial period, the sanctuary of Fortuna was enlarged and elaborated, the natural slope being shaped into a series of terraces linked by stairs.

Figure: *Steep ground at Powys Castle, North Wales, falls away in a series of terraces, some supported on vaulted undercrofts.*

The imperial villas at Capri were built to take advantage of varied terraces. At the seaside Villa of the Papyri in Herculaneum, the villa gardens of Julius Caesar's father-in-law fell away in a series of terraces, giving pleasant and varied views of the Bay of Naples. Only some of them have been excavated. At Villa of Livia, probably part of Livia Drusilla's dowry brought to the Julio-Claudian dynasty, rooms in the cryptoporticus beneath terracing were frescoed with trees in bloom and fruit.

Italian Renaissance

During the Italian Renaissance, the formalized, civilizing imprint of human control over wild nature expressed in terracing that was combined with stairs and water features, drew villa patrons and garden designers to escarpments that surveyed a handsome prospect. At the influential Cortile del Belvedere at the Vatican Palace, perfected under a series of popes from the earliest 16th century, the backdrop within the enclosed court was a raised terrace. The view in this case was from the Stanze of Raphael on an upper floor of the Palace.

English Landscape Garden

Even in the most naturalistic landscape gardens of Capability Brown, a raised gravelled or paved terrace along the garden front offered a dry walk in damp weather and a transition between the hard materials of the architecture and the rolling greensward beyond.

Contemporary

Contemporary terrace gardens, in addition to being in the garden and landscape, often occur in urban areas and are terrace architecture elements that extend out from an apartment or residence at any floor level other than ground level. They are often discussed in conjunction with roof gardens, although they are not always true roof gardens, instead being balconies and decks. These outdoor spaces can become lush gardens through the use of container gardening, automated drip irrigation and low-flow irrigation systems, and outdoor furnishings.

Green Belt

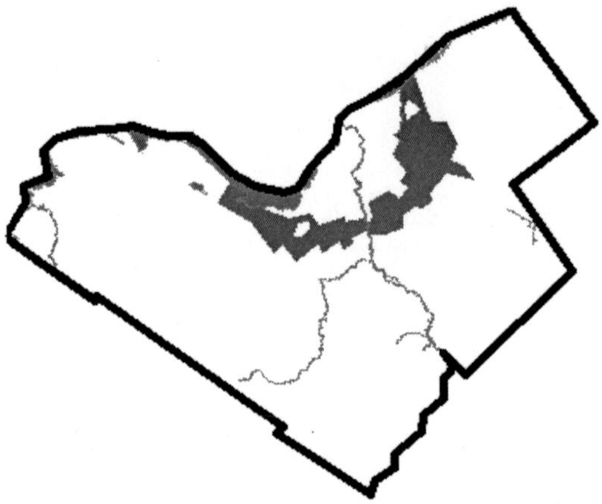

Figure: *The central core of Ottawa, located in the middle of the map, is surrounded by the Ottawa Greenbelt*

A green belt or greenbelt is a policy and land use designation used in land use planning to retain areas of largely undeveloped, wild, or agricultural land surrounding or neighbouring urban areas. Similar concepts are greenways or green wedges which have a linear character and may run through an urban area instead of around it. In essence, a green belt is an invisible line designating a border around a certain area, preventing development of the area and allowing wildlife to return and be established.

Purposes

In those countries which have them, the stated objectives of green belt policy are to:

- Protect natural or semi-natural environments;
- Improve air quality within urban areas;
- Ensure that urban dwellers have access to countryside, with consequent educational and recreational opportunities; and

- Protect the unique character of rural communities that might otherwise be absorbed by expanding suburbs.

The green belt has many benefits for people:

- Walking, camping, and biking areas close to the cities and towns.
- Contiguous habitat network for wild plants, animals and wildlife.
- Cleaner air and water
- Better land use of areas within the bordering cities.

The effectiveness of green belts differs depending on location and country. They can often be eroded by urban rural fringe uses and sometimes, development 'jumps' over the green belt area, resulting in the creation of "satellite towns" which, although separated from the city by green belt, function more like suburbs than independent communities.

History

The Old Testament outlines a proposal for a green belt around the Levite towns in the Land of Israel Moses Maimonides expounded that the greenbelt plan from the Old Testament referred to all towns in ancient Israel. In the 7th century, Muhammad established a green belt around Medina. He did this by prohibiting any further removal of trees in a 12-mile long strip around the city. In 1580 Elizabeth I of England banned new building in a 3-mile wide belt around the City of London in an attempt to stop the spread of plague. However, it was possible to buy dispensations which reduced the effectiveness of the proclamation.

In modern times, green belt policy was pioneered in the United Kingdom in the 1930s after pressure from the CPRE and various other organizations. There are fourteen green belt areas, in the UK covering 16,716 km^2, or 13% of England, and 164 km^2 of Scotland; for a detailed discussion of these. Other notable examples are the Ottawa Greenbelt and Golden Horseshoe Greenbelt in Ontario, Canada. Ottawa's 20,350 hectare greenbelt is managed by the National Capital Commission (NCC). The more general term in the U.S. is green space or greenspace, which may be a very small area such as a park.

The concept of "green belt" has evolved in recent years to encompass not only "Greenspace" but also "Greenstructure", taking into account all urban greenspaces, an important aspect of sustainable development in the 21st century. The European Commission's COST Action C11 (COST - European Cooperation in Science and Technology) is undertaking "Case studies in Greenstructure Planning" involving 15 European countries.

An act of the Swedish parliament from 1994 has declared a series of parks in Stockholm and the adjacent municipality of Solna to its north a "national city park" called Ekoparken (the "Eco park"; it stretches from the parks surrounding the royal palaces of Ulriksdal and Haga in Solna, through the Brunnsviken area, down to the former royal hunting grounds of North and South Djurgården).

Criticism

The difference/contrarian interpretation of the green belt's effects/ motivation (for example, suggested by economist Tim Harford) is that a green belt is created by residents to preserve the bourgeois status quo of those already living within the zone, and especially the advantage of landlords who profit from a scarcity of housing. In this interpretation, the stated motivation and benefits of the green belt are well-intentioned (public health, environment), but these benefits accrue as intentioned or claimed (for example, critics such as Mark Pennington claim that only a small fraction of the population ever sets foot on the green belt for leisure purposes, and they claim that a green belt is not strongly causally linked to clean air and water). Rather, the ultimate result of the decision to green belt a city is to maintain the middle class status quo, thus exacerbating high housing prices by concentrating demand within the zone and stifling competitive forces in general.

Another area of criticism comes from the fact that, since a greenbelt does not extend indefinitely outside a city, it might spur the growth of areas much further away from the city core than if it had not existed, thereby actually *increasing* urban sprawl. Examples commonly cited are the Ottawa suburbs of Kanata and Orleans, both of which are outside the city's greenbelt, and are currently undergoing explosive growth). This can lead to other problems, as residents of these areas have further to commute to work (if they seek employment in city) and little access to public transport. It also means people will commute through the green belt, an area not designed to cope with high levels of transportation. Not only is the merit of a green belt apparently subverted, but the green belt may heighten the problem and make the city unsustainable.

There are many examples whereby the actual effect of green belts is to act as a land reserve for future freeways and other highways. Examples include sections of the 407 highway north of Toronto and the Hunt Club Rd / Richmond Rd. south of Ottawa. Whether they are originally planned as such, or the result of a newer administration taking advantage of land that was left available by its predecessors is debatable.

In Britain, greenbelt barriers to urban expansion have been criticised as one of several protectionist political-economic barriers to housebuilding with negative effects on the supply, cost/prices, and quality of new homes. Critics argue that the greenbelts actually defeat their own stated objective of saving the countryside and open spaces. By preventing existing towns and cities from extending normally and organically, they result in more land-extensive housing developments further out – i.e., the establishment beyond the greenbelts of new communities with lower building densities, their own built infrastructure and other facilities, and greater dependence on cars and commuting, etc. Meanwhile, valuable urban green space and brownfield sites best suited to industry and commerce are lost in existing conurbations as more and more new housing is crammed into them.

Notable Examples

Australia:

- Adelaide's Central Business District is completely encircled by the Adelaide Parklands, as was initially planned in 1837.
- The Nillumbik Shire Council which is located approximately 30 km (19 miles) north-east of Melbourne is considered as "The Green Wedge Shire" because of the agreement with the Victorian Government which prevents high-density infrastructure to be built.

Brazil:

- The São Paulo City Green Belt Biosphere Reserve - GBBR, an integral part of the Atlantic Forest Biosphere Reserve, was created in 1994 stemming from a people's movement that collected 150 thousand signatures. It extends throughout 73 municipalities including São Paulo metro and the Santos area. With approximately 17,000 km^2, it is inhabited by about 23 million people, corresponding to more than 10% of the country's total population in an area equivalent to 2 thousandth of the Brazilian territory. There are over 6,000 km^2 of forests and other Atlantic Forest ecosystems at the Reserve, one of the planet's most threatened biomes. In addition to a spectacular biological diversity, the GBBR's ecosystems render valuable ecosystem Services.

Canada:

- Ottawa Greenbelt - Surrounds the Capital city of Ottawa
- Greenbelt (Golden Horseshoe), is a 7300 km^2 band of land that encompasses the rural and agricultural land surrounding the Greater Toronto Area and Niagara Peninsula, and parts of the

Bruce Peninsula. It's the largest greenbelt in the world and larger than Canada's smallest province - PEI. Most of the land consists of the Oak Ridges Moraine, an environmentally sensitive land that is a major aquifer for the region, and the Niagara Escarpment, a UNESCO Biosphere Reserve. In an effort to restrain urban sprawl that has encroched on these lands, the Ontario government created the Greenbelt Act in February 2005 to protect this greenspace from all future development, with the exception of limited agricultural use.

- British Columbia - the Agricultural Land Reserve protects agricultural land throughout this mountainous province from urban development, including around Vancouver. This protection is strict and urban development of agricultural land is only allowed if no reasonable alternative exists. However, it does not protect non-agricultural land, particularly hillsides, leading to substantial, and highly visible, leapfrog-type hillside sprawl.
- Quebec - The *Commission de protection du territoire agricole du Québec* asserts its mission, namely to keep a territory (the agricultural zones) that is favourable for the practice and the development of agricultural activities. In so doing, the commission safeguards the agricultural territory and helps make its protection a local priority. The agricultural zones cover an area of 63 000 square kilometres in 952 local municipalities.

Dominican Republic:

- The Greater Santo Domingo has a Greenbelt (Santo Domingo Greenbelt) project surrounding the whole Distrito Nacional. It is composed of the National botanical Garden, Mirador Del Norte, Mirador del Este, and other parks surrounding the area from its outer *municipios*. It has largely been affected by uncontrolled urbanization, but other parts remain unaffected.

Mainland Europe:

- European Green Belt
- Banjica Forest, Belgrade
- Stockholm Eco park
- German Green Belt
- Coulée verte du sud parisien
- Coulée verte du nord parisien
- Promenade plantée
- Vienna Woods, Austria

New Zealand:

- Dunedin's Town Belt is one of the world's oldest green belts, having been planned at the time of the city's rapid growth during the Otago Gold Rush of the 1860s. It surrounds the city centre on three sides (the fourth side being the city's harbour).

Pakistan:

- Islamabad, often called the "green city," is known for its green belts found on most roadsides which are often decorated and filled with various flora.

Philippines:

- Makati City's green belt is very green yet full of malls and modern structures.

Thailand:

- Bangkok's Bang Krachao Green Area located inside the curve of Chao Phraya River is considered a green area with authority control over the urbanization. Today it is a popular spot for tourism and cycling. The area is located within the boarder of Bangkok Province and Samut Sakorn Province.

South Korea:

- Seoul

United ***Kingdom:***

- The Metropolitan Green Belt (5,133 km^2)
- The North West Green Belt (2,578 km^2)
- South and West Yorkshire Green Belt (2,556 km^2)
- West Midlands Green Belt (2,315 km^2)

United ***States:***

- The U.S. states of Oregon, Washington and Tennessee require cities to establish urban growth boundaries (UGBs).
- Notable US cities which have adopted UGBs include Portland, Oregon; Twin Cities, Minnesota; Virginia Beach, Virginia; Lexington, Kentucky; and Miami-Dade County, Florida.
- More than 20 cities in the San Francisco Bay Area have UGBs
- Staten Island Greenbelt and Brooklyn-Queens Greenway in New York City
- Barton Creek Greenbelt, Austin
- Ann Arbor, Michigan is acquiring conservation easements on agricultural land around the city without the establishment of

an urban growth boundary. While the city's initial plan did not include the participation of surrounding townships, at least four townships have participated directly or have initiated their own efforts to protect agricultural land surrounding the city.

- Boise Greenbelt, Boise, Idaho
- *The Jungle*, Seattle
- The Emerald Necklace in Boston is halfway between a green belt and a greenway, nearly ringing central Boston. The final link in the chain, the Dorchesterway, was never constructed.

Plant Nutrition

Plant nutrition is the study of the chemical elements and compounds that are necessary for plant growth, and also of their external supply and internal metabolism. In 1972, E. Epstein defined two criteria for an element to be essential for plant growth:

1. in its absence the plant is unable to complete a normal life cycle; or
2. that the element is part of some essential plant constituent or metabolite.

This is in accordance with Liebig's law of the minimum. There are 14 essential plant nutrients. Carbon and oxygen are absorbed from the air, while other nutrients including water are obtained from the soil. Plants must obtain the following mineral nutrients from the growing media:

- the primary macronutrients: nitrogen (N), phosphorus (P), potassium (K)
- the three secondary macronutrients: calcium (Ca), sulphur (S), magnesium (Mg)
- the micronutrients/trace minerals: boron (B), chlorine (Cl), manganese (Mn), iron (Fe), zinc (Zn), copper (Cu), molybdenum (Mo), nickel (Ni)

The macronutrients are consumed in larger quantities and are present in plant tissue in quantities from 0.2% to 4.0% (on a dry matter weight basis). Micro nutrients are present in plant tissue in quantities measured in parts per million, ranging from 5 to 200 ppm, or less than 0.02% dry weight.

Most soil conditions across the world can provide plants with adequate nutrition and do not require fertilizer for a complete life cycle. However, humans can artificially modify soil through the addition of fertilizer to promote vigorous growth and increase yield. The plants are able to obtain their required nutrients from the fertilizer added to the soil. A colloidal carbonaceous residue, known as humus, can serve as a nutrient reservoir. Even with adequate water and sunshine, nutrient deficiency can limit growth.

Nutrient uptake in the soil is achieved by cation exchange, where root hairs pump hydrogen ions (H^+) into the soil through proton pumps. These hydrogen ions displace cations attached to negatively charged soil particles so that the cations are available for uptake by the root.

Plant nutrition is a difficult subject to understand completely, partly because of the variation between different plants and even between different species or individuals of a given clone. An element present at a low level may cause deficiency symptoms, while the same element at a higher level may cause toxicity. Further, deficiency of one element may present as symptoms of toxicity from another element. An abundance of one nutrient may cause a deficiency of another nutrient. For example, lower availability of a given nutrient such as SO_2^{4-} can affect the uptake of another nutrient, such as NO_3^-. As another example, K^+ uptake can be influenced by the amount of NH_4^+ available.

The root, especially the root hair, is the most essential organ for the uptake of nutrients. The structure and architecture of the root can alter the rate of nutrient uptake. Nutrient ions are transported to the centre of the root, the stele in order for the nutrients to reach the conducting tissues, xylem and phloem. The Casparian strip, a cell wall outside of the stele but within the root, prevents passive flow of water and nutrients, helping to regulate the uptake of nutrients and water. Xylem moves water and inorganic molecules within the plant and phloem accounts for organic molecule transportation. Water potential plays a key role in a plants nutrient uptake. If the water potential is more negative within the plant than the surrounding soils, the nutrients will move from the region of higher solute concentration—in the soil—to the area of lower solute concentration: in the plant.

There are three fundamental ways plants uptake nutrients through the root:

1) Simple diffusion, occurs when a nonpolar molecule, such as O_2, CO_2, and NH_3 follows a concentration gradient, moving passively through the cell lipid bilayer membrane without the use of transport proteins.
2) Facilitated diffusion, is the rapid movement of solutes or ions following a concentration gradient, facilitated by transport proteins.

3) Active transport, is the uptake by cells of ions or molecules against a concentration gradient; this requires an energy source, usually ATP, to power molecular pumps that move the ions or molecules through the membrane.

- Nutrients are moved inside a plant to where they are most needed. For example, a plant will try to supply more nutrients to its younger leaves than to its older ones. When nutrients are mobile, symptoms of any deficiency become apparent first on the older leaves. However, not all nutrients are equally mobile. Nitrogen, phosphorus, and potassium are mobile nutrients, while the others have varying degrees of mobility. When a less mobile nutrient is deficient, the younger leaves suffer because the nutrient does not move up to them but stays in the older leaves. This phenomenon is helpful in determining which nutrients a plant may be lacking.

Many plants engage in symbiosis with microorganisms. Two important types of these relationship are

1) with bacteria such as rhizobia, that carry out biological nitrogen fixation, in which atmospheric nitrogen (N_2) is converted into ammonium (NH_4); and
2) with mycorrhizal fungi, which through their association with the plant roots help to create a larger effective root surface area. Both of these mutualistic relationships enhance nutrient uptake.

Though nitrogen is plentiful in the Earth's atmosphere, relatively few plants harbor nitrogen fixing bacteria, so most plants rely on nitrogen compounds present in the soil to support their growth. These can either be supplied by decaying matter, nitrogen fixing bacteria, animal waste, or through the agricultural application of purpose-made fertilizers.

Hydroponics, is a method for growing plants in a water-nutrient solution without the use of nutrient-rich soil. It allows researchers and home gardeners to grow their plants in a controlled environment. The most common solution, is the Hoagland solution, developed by D. R. Hoagland in 1933, the solution consists of all the essential nutrients in the correct proportions necessary for most plant growth. An aerator is used to prevent an anoxic event or hypoxia. Hypoxia can affect nutrient uptake of a plant because without oxygen present, respiration becomes inhibited within the root cells. The Nutrient film technique is a variation of hydroponic technique. The roots are not fully submerged, which allows for adequate aeration of the roots, while a "film" thin layer of nutrient rich water is pumped through the system to provide nutrients and water to the plant.

Processes

Plants take up essential elements from the soil through their roots and from the air (mainly consisting of nitrogen and oxygen) through their leaves. Nutrient uptake in the soil is achieved by cation exchange, wherein root hairs pump hydrogen ions (H+) into the soil through proton pumps. These hydrogen ions displace cations attached to negatively charged soil particles so that the cations are available for uptake by the root. In the leaves, stomata open to take in carbon dioxide and expel oxygen. The carbon dioxide molecules are used as the carbon source in photosynthesis.

Functions of Nutrients

Although nitrogen is plentiful in the Earth's atmosphere, relatively few plants engage in nitrogen fixation (conversion of atmospheric nitrogen to a biologically useful form). Most plants therefore require nitrogen compounds to be present in the soil in which they grow.

Plant nutrition is a difficult subject to understand completely, partially because of the variation between different plants and even between different species or individuals of a given clone. Elements present at low levels may cause deficiency symptoms, and toxicity is possible at levels that are too high. Furthermore, deficiency of one element may present as symptoms of toxicity from another element, and vice versa.

Carbon and oxygen are absorbed from the air, while other nutrients are absorbed from the soil. Green plants obtain their carbohydrate supply from the carbon dioxide in the air by the process of photosynthesis. Each of these nutrients is used in a different place for a different essential function.

Macronutrients (Derived from Air and Water)

Carbon: Carbon forms the backbone of many plants biomolecules, including starches and cellulose. Carbon is fixed through photosynthesis from the carbon dioxide in the air and is a part of the carbohydrates that store energy in the plant.

Hydrogen

Hydrogen also is necessary for building sugars and building the plant. It is obtained almost entirely from water. Hydrogen ions are imperative for a proton gradient to help drive the electron transport chain in photosynthesis and for respiration.

Oxygen

Oxygen by itself or in the molecules of H_2O or CO_2 are necessary for plant cellular respiration. Cellular respiration is the process of generating energy-rich adenosine triphosphate (ATP) via the consumption of sugars made in photosynthesis. Plants produce oxygen gas during photosynthesis to produce glucose but then require oxygen to undergo aerobic cellular respiration and break down this glucose and produce ATP.

Macronutrients (Primary)

Phosphorus: Phosphorus is important in plant bioenergetics. As a component of ATP, phosphorus is needed for the conversion of light energy to chemical energy (ATP) during photosynthesis. Phosphorus can also be used to modify the activity of various enzymes by phosphorylation, and can be used for cell signaling. Since ATP can be used for the biosynthesis of many plant biomolecules, phosphorus is important for plant growth and flower/seed formation. Phosphate esters make up DNA, RNA, and phospholipids. Most common in the form of polyprotic phosphoric acid (H_3PO_4) in soil, but it is taken up most readily in the form of H_2PO_4. Phosphorus is limited in most soils because it is released very slowly from insoluble phosphates. Under most environmental conditions it is the limiting element because of its small concentration in soil and high demand by plants and microorganisms. Plants can increase phosphorus uptake by a mutualism with mycorrhiza. A Phosphorus deficiency in plants is characterized by an intense green colouration in leaves. If the plant is experiencing high phosphorus deficiencies the leaves may become denatured and show signs of necrosis. Occasionally the leaves may appear purple from an accumulation of anthocyanin. Because phosphorus is a mobile nutrient, older leaves will show the first signs of deficiency.

It is useful to apply a high phosphorus content fertilizer, such as bone meal, to perennials to help with successful root formation.

Potassium: Potassium regulates the opening and closing of the stomata by a potassium ion pump. Since stomata are important in water regulation, potassium reduces water loss from the leaves and increases drought tolerance. Potassium deficiency may cause necrosis or interveinal chlorosis. K^+ is highly mobile and can aid in balancing the anion charges within the plant. Potassium helps in fruit colouration, shape and also increases its brix. Hence, quality fruits are produced in Potassium rich soils. It also has high solubility in water and leaches out of rocky or sandy soils. This water solubility can result in potassium deficiency. Potassium serves as an activator of enzymes used in photosynthesis and respiration Potassium is used to build cellulose and aids in photosynthesis by the formation of a chlorophyll precursor. Potassium deficiency may result in higher risk of pathogens, wilting, chlorosis, brown spotting, and higher chances of damage from frost and heat.

Nitrogen: Nitrogen is an essential component of all proteins. Nitrogen deficiency most often results in stunted growth, slow growth, and chlorosis. Nitrogen deficient plants will also exhibit a purple appearance on the stems, petioles and underside of leaves from an accumulation of anthocyanin pigments.

Most of the nitrogen taken up by plants is from the soil in the forms of NO_3^-, although in acid environments such as boreal forests where nitrification is less likely to occur, ammonium NH_4^+ is more likely to be the dominating source of nitrogen. Amino acids and proteins can only be built from NH_4^+ so NO_3^- must be reduced. Under many agricultural settings, nitrogen is the limiting nutrient of high growth. Some plants require more nitrogen than others, such as corn (*Zea mays*). Because nitrogen is mobile, the older leaves exhibit chlorosis and necrosis earlier than the younger leaves. Soluble forms of nitrogen are transported as amines and amides.

Macronutrients (Secondary and Tertiary)

Sulphur: Sulphur is a structural component of some amino acids and vitamins, and is essential in the manufacturing of chloroplasts. Sulphur is also found in the Iron Sulphur complexes of the electron transport chains in photosynthesis. It is immobile and deficiency therefore affects younger tissues first. Symptoms of deficiency include yellowing of leaves and stunted growth.

Calcium: Calcium regulates transport of other nutrients into the plant and is also involved in the activation of certain plant enzymes. Calcium deficiency results in stunting. This nutrient is involved in photosynthesis and plant structure. Blossom end rot is also a result of inadequate calcium.

Magnesium: Magnesium is an important part of chlorophyll, a critical plant pigment important in photosynthesis. It is important in the production of ATP through its role as an enzyme cofactor. Magnesium deficiency can result in interveinal chlorosis.

Silicon: In plants, silicon strengthens cell walls, improving plant strength, health, and productivity. Other benefits of silicon to plants include improved drought and frost resistance, decreased lodging potential and boosting the plant's natural pest and disease fighting systems. Silicon has also been shown to improve plant vigor and physiology by improving root mass and density, and increasing above ground plant biomass and crop yields. Although not considered an essential element for plant growth and development (except for specific plant species - sugarcane and members of the horsetail family), silicon is considered a beneficial element in many countries throughout the world due to its many benefits to numerous plant species when under abiotic or biotic stresses. Silicon is currently under consideration by the Association of American Plant Food Control Officials (AAPFCO) for elevation to the status of a "plant beneficial substance".

Silicon is the second most abundant element in earth's crust. Higher plants differ characteristically in their capacity to take up silicon. Depending on their SiO_2 content they can be divided into three major groups:

- Wetland graminae-wetland rice, horsetail (10–15%)
- Dryland graminae-sugar cane, most of the cereal species and few dicotyledons species (1–3%)
- Most of dicotyledons especially legumes (<0.5%)
- The long distance transport of Si in plants is confined to the xylem. Its distribution within the shoot organ is therefore determined by transpiration rate in the organs
- The epidermal cell walls are impregnated with a film layer of silicon and effective barrier against water loss, cuticular transpiration rate in the organs.

Si can stimulate growth and yield by several indirect actions. These include decreasing mutual shading by improving leaf erectness, decreasing susceptibility to lodging, preventing Mn and Fe toxicity.

Micro-nutrients

Some elements are directly involved in plant metabolism (Arnon and Stout, 1939). However, this principle does not account for the so-called beneficial elements, whose presence, while not required, has clear positive effects on plant growth. Mineral elements that either stimulate growth but are not essential, or that are essential only for certain plant species, or under given conditions, are usually defined as beneficial elements.

Iron: Iron is necessary for photosynthesis and is present as an enzyme cofactor in plants. Iron deficiency can result in interveinal chlorosis and necrosis. Iron is not the structural part of chlorophyll but very much essential for its synthesis. Copper deficiency can be responsible for promoting an iron deficiency.

Molybdenum: Molybdenum is a cofactor to enzymes important in building amino acids. Involved in Nitrogen metabolism. Mo is part of Nitrate reductase enzyme.

Boron: Boron is important for binding of pectins in the RGII region of the primary cell wall, secondary roles may be in sugar transport, cell division, and synthesizing certain enzymes. Boron deficiency causes necrosis in young leaves and stunting.

Copper: Copper is important for photosynthesis. Symptoms for copper deficiency include chlorosis. Involved in many enzyme processes. Necessary for proper photosythesis. Involved in the manufacture of lignin (cell walls). Involved in grain production. It is also hard to find in some conditions.

Manganese: Manganese is necessary for photosynthesis, including the building of chloroplasts. Manganese deficiency may result in colouration abnormalities, such as discoloured spots on the foliage.

Sodium: Sodium is involved in the regeneration of phosphoenolpyruvate in CAM and C4 plants. It can also substitute for potassium in some circumstances.

Essentiality

- Essential for C4 plants rather C3
- Substitution of K by Na: Plants can be classified into four groups:
  1. Group A—a high proportion of K can be replaced by Na and stimulate the growth, which cannot be achieved by the application of K
  2. Group B—specific growth responses to Na are observed but they are much less distinct
  3. Group C—Only minor substitution is possible and Na has no effect
  4. Group D—No substitution is occurred
- Stimulate the growth—increase leaf area, stomata, improve the water balance
- Na functions in metabolism
  1. C4 metabolism
  2. Impair the conversion of pyruvate to phosphoenol-pyruva

   3. Reduce the photosystem II activity and ultrastructural changes in mesophyll chloroplast
- Replacing K functions
   1. Internal osmoticum
   2. Stomatal function
   3. Photosynthesis
   4. Counteraction in long distance transport
   5. Enzyme activation
- Improves the crop quality e.g. improve the taste of carrots by increasing sucrose

Zinc: Zinc is required in a large number of enzymes and plays an essential role in DNA transcription. A typical symptom of zinc deficiency is the stunted growth of leaves, commonly known as "little leaf" and is caused by the oxidative degradation of the growth hormone auxin.

Nickel: In higher plants, Nickel is absorbed by plants in the form of Ni^{2+} ion. Nickel is essential for activation of urease, an enzyme involved with nitrogen metabolism that is required to process urea. Without Nickel, toxic levels of urea accumulate, leading to the formation of necrotic lesions. In lower plants, Nickel activates several enzymes involved in a variety of processes, and can substitute for Zinc and Iron as a cofactor in some enzymes.

Chlorine: Chlorine, as compounded chloride, is necessary for osmosis and ionic balance; it also plays a role in photosynthesis.

Cobalt: Cobalt has proven to be beneficial to at least some plants, but is essential in others, such as legumes where it is required for nitrogen fixation for the symbiotic relationship it has with nitrogen-fixing bacteria. Vanadium may be required by some plants, but at very low concentrations. It may also be substituting for molybdenum. Selenium and sodium may also be beneficial. Sodium can replace potassium's regulation of stomatal opening and closing.

1. The requirement of Co for N_2 fixation in legumes and non-legumes have been documented clearly
2. Protein synthesis of Rhizobium is impaired due to Co deficiency
3. It is still not clear whether Co has direct effect on higher plant

Aluminium:

- Tea has a high tolerance for Al toxicity and the growth is stimulated by Al application. The possible reason is the prevention of Cu, Mn or P toxicity effects.

- There have been reports that Al may serve as fungicide against certain types of root rot.

Xeriscaping

Xeriscaping (less commonly known as xeroscape) is landscaping and gardening that reduces or eliminates the need for supplemental water from irrigation. It is promoted in regions that do not have easily accessible, plentiful, or reliable supplies of fresh water, and is gaining acceptance in other areas as water becomes more limiting. Xeriscaping may be an alternative to various types of traditional gardening.

In some areas, terms such as water-conserving landscapes, *drought-tolerant* landscaping, and *smart scaping* are used instead. Plants whose natural requirements are appropriate to the local climate are emphasized, and care is taken to avoid losing water to evaporation and run-off. The specific plants used in xeriscaping depend upon the climate. Xeriscaping is different from natural landscaping, because the emphasis in xeriscaping is on selection of plants for water conservation, not necessarily selecting native plants.

Advantages

Figure: *Cacti are one of the low-water-consuming plants used in Xeriscaping.*

- Lowers consumption of imported or ground water.
- More water available for other domestic and community uses and the environment.
- Less time and work needed for maintenance effort (no lawns to cut).
- Xeriscape plants in appropriate planting design, and soil grading and mulching, takes full advantage of rainfall retention.

- When water restrictions are implemented, by municipality or water costs, xeriscape plants will tend to survive and thrive, while more ornamental plants may be unable to adapt.
- Can be visually more interesting than lawns.
- Saves a lot of money
- Sometimes deemed aesthetically unpleasant by traditionalists.

Disadvantages

- Compared to grass which is soft, dry landscapes often make use of gravel, rocks or woodchips which can hurt or cut bare feet.
- Landscape may be considerably less useful for outdoor activities and sports than a lawn.
- Certain plants such as cacti and agave contain thorns or serrated edges which may harm pets and children.
- May be expensive to plan.
- Extensive gardening in otherwise dry climates may host conditions for wildfires. (ex: Eucalyptus grows in arid Australia, but produces highly flammable oils and resins).
- May promote possible reintroduction of unwanted native species (Black widows, rattlesnakes, Coyotes), however existing heavily watered landscapes act as oasis to such species anyway.

X-ratings

In 1982 the Denver water Department's Xeriscape Demonstration Garden was dedicated with an "X-rated party". The X stood for Xeriscape, but the humor was intended. Garden centres continued with a system of X-ratings, from X to XXX, to specify how much water individual types of plants need.

Denver Water Summary

The Xeriscape concept has been denoted by Denver Water in 7 basic principles:

Plan and design – Create a diagram, drawn to scale, that shows the major elements of the landscape, including house, driveway, sidewalk, deck or patio, existing trees and other elements.

Once a base plan of an existing site has been determined, the creation of a conceptual plan (bubble diagram) that shows the areas for turf, perennial beds, views, screens, slopes, etc. is undertaken. Once finished, the development of a planting plan that reinforces the areas in the appropriate scale is done.

Soil amendment – Most plants will benefit from the use of compost, which will help the soil retain water. Some desert plants prefer gravel soils instead of well-amended soils. Plants should either fit the soil or soil should be amended to fit the plants.

Efficient irrigation – Xeriscape can be irrigated efficiently by hand or with an automatic sprinkler system. Zone turf areas separately from other plants and use the irrigation method that waters the plants in each area most efficiently. For grass, use gear-driven rotors or rotary spray nozzles that have larger droplets and low angles to avoid wind drift. Spray, drip line or bubbler emitters are most efficient for watering trees, shrubs, flowers and groundcovers.

If watering by hand, avoid oscillating sprinklers and other sprinklers that throw water high in the air or release a fine mist. The most efficient sprinklers release big drops close to the ground.

Water deeply and infrequently to develop deep roots. Never water during the day to reduce water lost to evaporation. With the use of automatic sprinkling systems, adjust the controller monthly to accommodate weather conditions. Also, install a rain sensor to shut off the device when it rains.

Appropriate plant and zone selection – Different areas in a yard receive different amounts of light, wind and moisture. To minimize water waste, group together plants with similar light and water requirements, and place them in an area that matches these requirements. Put moderate-water-use plants in low-lying drainage areas, near downspouts, or in the shade of other plants. Turf typically requires the most water and shrub/perennial beds will require approximately half the amount of water. Dry, sunny areas support low-water-use plants that grow well in our climate. Planting a variety of plants with different heights, colour and textures creates interest and beauty.

Mulch – Mulch keeps plant roots cool, prevents soil from crusting, minimizes evaporation and reduces weed growth. Organic mulches, such as bark chips, pole peelings or wood grindings, should be applied 2 to 4 inches deep. Fibre mulches create a web that is more resistant to wind and rain washout. Inorganic mulches, such as rocks and gravel, should be applied 2 to 3 inches deep. Surrounding plants with rock makes the area hotter; limit this practice.

Alternative turf – Native grasses (warm-season) that have been cultivated for turf lawns, such as buffalo grass and blue grama, can survive with a quarter of the water that bluegrass varieties need.

Warm-season grasses are greenest in June through September and straw brown the rest of the year.

Native grasses (cool season) such as bluegrass and tall fescue, are greenest in the spring and fall and go dormant in the high heat of the summer. New cultivars of bluegrass, such as Reveille, and tall fescue, can reduce typical bluegrass water requirements by at least 30 percent. Fine fescues can provide substantial water savings and is best used in areas that receive low traffic or are in shady locations.

Use the appropriate grass and limit the amount of grass to reduce the watering and maintenance requirements.

Maintenance – All landscapes require some degree of care during the year. Turf requires spring and fall aeration along with regular fertilization every 6 to 8 weeks. Keep the grass height at 3 inches and allow the clippings to fall. Trees, shrubs and perennials will need occasional pruning to remove dead stems, promote blooming or control height and spread. Much of the removed plant material can be shredded and used in composting piles.

Physiological Responses of Plants to Phenolic Compounds

Various compounds of phenols are common in plant tissues, each of which may enclose a number of grams per every kilogram. Phenol compounds present in plant tissues begin to blend as a result of stimulants on the outer surface like ultraviolet radiation, microbial infections or and chemical changes. Most phenol amalgams are phytochemicals that are basically found in some plant tissues and there are as many as 4000 such compounds.

All plants create numerous organic amalgams that are always not related to the basic metabolism like progress, growth and development and the functions of these natural compounds in plants have only been noticed recently in a diagnostic perspective. Although these organic amalgam produced by the plants do not help in growth, they are effective in other ways. While some of these protect plants from predators and pathogens, others help in the reproductive process by fascinating pollinating agents or scattering the seeds. Many of them are also useful as they help to produce poisons that protect the plants.

Most of the organic products found in plants may be grossly put into three broad categories – alkaloids, phenylpropanoids and terpenoids. While phenylpropanoids are mostly phenolic amalgams, terpenoids are made up of five carbon elements that are blended through the acetate or mevalonate method or the glyceraldehydes 3-phospate/pyruvate process. Many of the terpenoids produced by the plants are a sort of contaminants that protect the plants from being devoured by herbivores creatures or functionally attractants for pests and animals.

Alkaloids, containing nitrogen amalgams, are mainly blended from amino acids and also protect the plants from herbivorous animals and may also be useful pharmacologically. On the other hand, phenolic amalgams are created by through fusion of shikimic acid products and have different functions in plants. Some of the common phenolic amalgams or compounds include tannins, lignans and flavonoids that protect plants from herbivores and other destroyers.

Basically, all phenols and phenolic amalgams are widely found in nature and can also be blended artificially. They form a separate group of chemical substances that comprise a member of hydroxyl cluster linked to an element of hydrocarbon set. Commonly found in substances like antiseptics, resins, plastics, cosmetics, foods and beverages, health aid as well as medicines, these amalgams have varied industrial uses. Many of them like pentachlorophenol find uses in pesticides as well as preservatives for building items, especially wood and rubber. Incidentally, they are grave contaminants and need to be handled carefully as they may contaminate the ground water or pollute the atmosphere during industrial use. Hence, there are several regulations restricting the use of these compounds.

A number of vital vegetation ingredients including tannins, phenolic amalgams, flavones, coumarins, anthraquinones and all their glycosides are comprise phenolic molecules.

Simple Phenols

Phenols particles are created when the hydrogen content in a pungent nucleus is swapped with an element from the hydroxyl cluster and this process may happen several times making the isomers include three probable dihydroxylbenzenes as well as three trihydroxylbenzenes. While the dihydroxylbenzenes are hydroquinone, catechol and resorcinol, the trihydroxylbenzenes include benzenitrol, pyrogallol and phloroglucinol. Interestingly, the hydroxyl set in phenol molecule imparts features of alcohol and tends to classify phenols as tertiary alcohols; basically the hydroxyl group in phenol is linked to a dilute carbon nucleus. Thus, phenols are not alcohols, but actually classified under weak acids.

Nearly all phenol byproducts of nearly all phenols are created from several circles and hence they are basically polycyclic molecules changing their basic features. When fundamental elements are replaced by one or more hydrogen, it leads to the formation on simple phenols. The material becomes more bioactive or toxic when these fundamental elements belong to alkyl groups like in eugenol or thymol. Basically, phenols are used to eliminate bacteria, as antiseptics and also as poisons to burn up parasites (anthelmintic). When the crust or the internal arrangement of the cells are deliberately damaged phenols have an antimicrobial effect, and like most antibacterial instruments, slow down leucocyte functioning and leads to crisis in antibiotics like penicillin. Plant tissues containing phenols comprise agents that stimulate leucocyte and recompense, the association of antimicrobial medicines along with substances that weaken the body's resistance power is always harmful.

Salicylic Acid

Although not identified specifically, salicylic acid or its properties first came to the light in the 5th century BC when the Greek physician Hippocrates mentioned about a bitter powder removed from willow bark that may perhaps cure pains and diminish fevers. Even the Sumer, Lebanon, and Assyria civilizations made use of this medicine. On the other hand, for years the Native Americans and Cherokee consumed a mixture prepared from the bark to cure fever and other ailments. In fact, the remedial part is found in the inner bark of willow and is generally used in curing pains and other diseases.

The medicinal extract of the bark later came to be known as salicin from Salix alba, which is white willow in Latin. French pharmacist Henri Leroux and Italian chemist Raffaele Piria were the first to extract the medicinal substance and store it in a crystalline form in 1828. Next, they transformed the substance into sugar and another element which after reaction with oxygen turned to salicylic acid (SA).

Ever present in plants, salicylic acid is basically a phytohormone and a phenol that is responsible for numerous functions, including photosynthesis, growth and maturity, transportation, transpiration, ion consumption and also brings about certain changes in the leaf structure and chloroplast formation. At the same time, owing to phenol content, salicylic acid also plays a vital role in plant defense by protecting it from pathogens and also helps in the manufacture of pathogenesis - related proteins that increase the resistance power in pathogens. Salicylic acid is obtained by bio synthesis of amino acid phenylalanine.

Salicylic acid, found widely as ester methyl salicylate in many plant groups, is the base for all medicines in the aspirin group. Salicylic acid imparts the active (aglycone) factor to glycosides like salicin from the willow (Salix spp.) and poplar (Populus spp.) groups, gaultherin from wintergreen (Gaultheria procumbens), birch (Betula spp.), and

Indian pipe (Monotropa uniflora) and from cramp bark (Viburnum opulus) and black haw bark (Viburnum prunifolium); populin from the poplar family. In 1839, German researchers also extracted salicylic acid from meadowsweet (Filipendula ulmaria, earlier categorized as Spiraea ulmaria). Although this extract was helpful in curing a number of ailments, it also had adverse side-effects like bleeding, diarrhea, gastric irritation and even resulted to death following over doses.

Interestingly, aspirin has derived it name from the previous botanical name of meadowsweet – Spiraea and salicylic acid got is name from the fact that it was first extracted from the bark of willow tree. As salicylic acid is a carboxylated phenol, it is a very sturdy antiseptic.

Contrary to salicylates, esters and salts are less vigorous and hence not as much of infuriating to the stomach lining and in fact, when acetylsalicylic acid was manufactured in bulk it was found to have little irritating effect. A vital function of the salicylates is to restrain the agents of swelling, irritation and prostaglandins that might otherwise even result to blood clotting in the body. Pharmacologically speaking, salicylates are very vigorous and when used externally they are feeble anodynes and cause irritation. However, small concentrations of salicylates found in some plants help in the renewal of epithelial tissues.

When used internally, salicylates help in creasing the flow of blood in the surrounding area and also increase sweat formation owing to they strong antipyretic nature. These are a result of an express exploit on the thermogenic hub in the hypothalamus. Salicylates enhance the body heat in to recompense the temperature loss in persons with pyrexia and also act on the liver cells (hepatocytes) by augmenting the amount and concentration of bile. In fact, salicylates are catabolic in nature and hence enhance sulphur and urinary urea secretion from the body and also augment uric acid concentration.

Salicylates are contra-indicated in gout, as along with the above functions, they also lead to foster uric acid discharge in the kidney and have a mild diuretic consequence. On the other hand, everyone is aware of the analgesic functions of salicylates and this is the primary result on the acuity of aches or might engage action on prostaglandin metabolism, or both.

In acidic conditions, byproducts of salicylates transform into salicylic acid offering useful antiseptic results that in areas of acute pains like arthritic joints and in urinary tubules. Recently, medicos are using the anti-clotting features of salicylates as a prophylactic healing for thrombotic state. As the issue is still being researched, much cannot be yet said about the usefulness of the plants containing salicylates in this regard. On the other hand, the oil of wintergreen that contains methyl salicylate is applied externally and is very useful as a rubefacient that quickly gets soaked up by the skin. This makes the substance a valuable medicine for curing inflamed joint as it not only diminishes the inflammatory procedure, but also effortlessly augments the flow of blood in the area.

Salicylates are known to have serious side-effects like generating numerous bleedings in the stomach particularly if they are used in a concentrated state. Although not so grave always, salicylates led to stomach hemorrhage in almost all people at some time or the other and in such cases meadowsweet extracts are the best to heal the damage. This is an irony indeed and is generated by the amalgamation of mucilages and tannins, which is normally abandoned during the conservative study of medicinal plant chemistry.

Many skin-care products use salicylic acid as the main constituent for curing acne, psoriasis, corns, calluses, warts and keratosis pilaris, as it can easily marsh the epidermis and check the pores from choking up and making room for new cells. Salicylic acid also forms an ingredient of many shampoos to cure dandruff and in gels to eliminate warts owing to its stimulation on epidermal cells. However, use of intense salicylic acid suspensions many be harmful for dark complexioned people and also for those who do not use sun blocks. These type of people are susceptible to hyper-pigmentation and un-pretreated skin.

Salicylates have other medicinal qualities too and since long have been used to get respite from fever and remedies for inflammatory diseases. To refresh the skin, salicylic acids may be used by drenched in cotton fillings. Sub-salicylate together with bismuth materializes into Pepto-Bismol, a commonly used medicine for tummy reprieve. Blended together the two ingredients not only form a moderate antibiotic, but also an effective remedy for nausea, diarrhea, gas and heartburn.

Eugenol

A member of the allybenzene category of chemical amalgams, eugenol is basically a guaiacol formed by replacing an allyl chain. Normally, eugenol is a light yellow slippery fluid obtained from vital oils like nutmeg, clove oil, bay leaf and cinnamon. Soluble to some extent in water and other organic fluids, eugenol smells pleasing, peppery and clove-like. Cloves, biologically called Syzygium aromaticum, syn. Eugenia aromaticum or Eugenia caryophyllata, are actually dried scented flower buds from a tree belonging to the Myrtaceae group that is commonly found in Indonesia, Zanzibar and Madagascar and is popularly exploited as a spice everywhere. Deriving its name from the French word 'clou' meaning a 'nail', cloves are also cultivated in India and Sri Lanka and they get their fragrance from the chemical substance called eugenol. Present in cloves in as high quantity as 72-90 percent, eugenol possesses distinct antiseptic and anesthetic qualities.

The substance has multiple uses, including in perfumeries, flavorings, edible oils and medicines. In medicines, eugenol is commonly used as local antiseptics and anesthetics. Although most of vanillin is produced artificially from phenol or lignan, eugenol is also widely used to produce isoeugenol for further manufacture of vanillin. Blending eugenol and zinc oxide results into zinc oxide eugenol that is used for interim filling and cementing decayed teeth. It also has antibacterial and anesthetic properties and is classified as an intermediary curative substance.

Interestingly, eugenol shows the normal anesthetic and antiseptic qualities of phenols and hence has been used to get relief from tooth aches and also used as antiseptic mouthwash. It also has all other qualities of clove oil and can be used externally as counter-irritant and internally to stimulate saliva discharge to control gastronomic functions. When used inside the intestines, the chemical functions as carminatives and also to avoid irregularities.

Thymol

Thymols, used as antiseptics, local anesthetics, preservatives as well as cooling agents, are outsized, pale and semi-transparent hexagonal sparklers or crystals which are perfumed like thyme and have a sharp and scented taste. Thymol is an ingredient used in lip creams, toothpaste, mouthwash as well as liniments and also locally applied as an irritant and anesthetic to skin and mucous membrane. The chemical does not match well with macrobiotic substances, oxidizing agents or strong bases and it is harmful if taken internally. Hence, swallowing, inhaling or allowing thymol to be soaked in the skin is dangerous. Thymol causes irritation when it comes in contact with eyes, skin and the respiratory system.

Obtained from the oil of thyme (Thymus vulgaris) and present in perfumed labiates, especially in horsemint (Monarda punctata), thymol is a useful antifungal negotiator. It is normally used at intensities of 1-2 percent (thyme oil contains 20-30 per cent). Thymol is also an effective medicine to get relief from oral and intestinal diseases and compared to phenols, it is 20 times stronger as an antiseptic. Thymol is also an efficient anthelmintic. Interestingly, although thymol is much more resilient than phenol, it has less mucosal aggravating qualities than any member of the cluster. Since thymol is rarely taken in by the body fluids, it is mainly used externally.

Hydroquinone

Hydroquinone is the aglycone of arbutin commonly present in the heather group – Ericaceae and Rosaceae. Interestingly, hydroquinone is full of arbutin and is usually applied as a urinary antiseptic, as arbutin is translated into hydroquinone in the urine. Like other members of the phenol family, hydroquinone has antiseptic properties especially against Klebsiella and Escherichia coli.

A topical skin lightening mediator, hydroquinone reduces the creation of melanin in the skin. Frequently described as a bleaching cream that eases pigment clusters, the effect of hydroquinone can be invalidated by rendering the skin to ultraviolet light. As the exposure of the sun can invalidate the action wrought by hydroquinone, a sunscreen is normally added to it during its manufacture. Hydroquinone is an effective medicine for curing murky skin, age spots, brown spots and melasma.

Hydroquinone is used to treat darkened skin areas which include melasma, freckles, age spots, and brown spots. Hydroquinone reduces manufacture of melanosomes (melanin pigment granules) in the skin's coloured cells by slowing down the functioning of tyrosinase, an enzyme required to produce melanin.

7

Plant Defenses against Pathogens and Herbivores

Plants represent a rich source of nutrients for many organisms including bacteria, fungi, protists, insects, and vertebrates. Although lacking an immune system comparable to animals, plants have developed a stunning array of structural, chemical, and protein-based defenses designed to detect invading organisms and stop them before they are able to cause extensive damage. Humans depend almost exclusively on plants for food, and plants provide many important non-food products including wood, dyes, textiles, medicines, cosmetics, soaps, rubber, plastics, inks, and industrial chemicals. Understanding how plants defend themselves from pathogens and herbivores is essential in order to protect our food supply and develop highly disease-resistant plant species.

This article introduces the concept of plant disease and provides an overview of some defense mechanisms common among higher plants. A close examination of plant anatomy is presented, as well as some of the ecological relationships that contribute to plant defense and disease resistance. Special care has been taken to illustrate how products used in everyday life are derived from substances produced by plants during defense responses.

Plant Disease and Resistance

Broadly defined, disease is any physiological abnormality or significant disruption in the "normal" health of a plant. Disease can be caused by living (biotic) agents, including fungi and bacteria, or by environmental (abiotic) factors such as nutrient deficiency, drought, lack of oxygen, excessive temperature, ultraviolet radiation, or pollution. In order to protect themselves from damage, plants have developed a wide variety of constitutive and inducible defenses. Constitutive (continuous) defenses include many preformed barriers such as cell walls, waxy epidermal cuticles, and bark. These substances not only protect the plant from invasion, they also give the plant strength and rigidity. In addition to preformed barriers, virtually all living plant cells have the ability to detect invading pathogens and respond with inducible defenses including the production of toxic chemicals, pathogen-degrading enzymes, and deliberate cell suicide. Plants often wait until pathogens are detected before producing toxic chemicals or defense-related proteins because of the high energy costs and nutrient requirements associated with their production and maintenance.

Plant Pathogens: Sneak Attack versus Brute Force

Many plant pathogens act like “silent thieves” who want to steal money locked inside of a bank vault. These thieves use specialized tools designed to disable the bank’s security system and unlock the vault without being detected. In a similar way, many pathogens establish intimate connections with their hosts in order to suppress plant defenses and promote the release of nutrients. Pathogens that keep their host alive and feed on living plant tissue are called biotrophs. Examples of biotrophic pathogens include the powdery mildew fungus *Blumeria graminis* and the bacterial rice pathogen *Xanthomonas oryzae*. Other pathogens resort to brute force like thieves who blast open a bank vault with explosives. These pathogens often produce toxins or tissue-degrading enzymes that overwhelm plant defenses and promote the quick release of nutrients. These pathogens are called necrotrophs, and examples include the gray mold fungus *Botrytis cinerea* and the bacterial soft-rot pathogen *Erwinia carotovora*. Some pathogens are biotrophic during the early stages of infection but become necrotrophic during the latter stages of disease. These pathogens are called hemibiotrophs and include the fungus *Magnaporthe grisea*, the causative agent of rice blast disease.

Most biotrophic and hemibiotrophic pathogens can only cause disease on a relatively small group of host plants because of the slightly different set of specialized genes and molecular mechanisms required for each host-pathogen interaction. The host range refers to the plant species on which a pathogen is capable of causing disease. For example, brome mosaic virus (BMV) infects grasses such as barley but not legumes. A plant species that does not show disease when infected with a pathogen is referred to as a non-host plant species for that pathogen. Organisms that do not cause disease on any plant species, such as the saprophytic bacterial species *Pseudomonas putida*, are referred to as non-pathogens.

When a pathogen is capable of causing disease on a particular host species, two outcomes are possible: A compatible response is an interaction that results in disease, while an incompatible response is an interaction that results in little or no disease at all. Although a particular plant species may be a susceptible host for a particular pathogen, some individuals may harbor genes that help recognize the presence of the pathogen and activate defenses. For example, some tomato cultivars show disease when infected with the bacterial pathogen *Pseudomonas syringae* (a compatible response), but others (cultivar Rio Grande, for example) are capable of recognizing the bacteria and limiting disease via resistance (an incompatible response). Disease resistance exists as a continuum of responses ranging from immunity (the complete lack of any disease symptoms) to highly resistant (some disease symptoms) to highly susceptible (significant disease symptoms).

Plants have developed multiple layers of sophisticated surveillance mechanisms that recognize potentially dangerous pathogens and rapidly respond before those organisms have a chance to cause serious damage. These surveillance systems are linked to specific pre-programmed defense responses. Basal resistance, also called innate immunity, is the first line of pre-formed and inducible defenses that protect plants against entire groups of pathogens. Basal resistance can be triggered when plant cells recognize microbe-associated molecular patterns (MAMPs) including specific proteins, lipopolysaccharides, and cell wall components commonly found in microbes. The result is that living plant cells become fortified against attack. Non-pathogens as well as pathogens are capable of triggering basal resistance in plants due to the widespread presence of these molecular components in their cells.

Pathogens have developed countermeasures that are able to suppress basal resistance in certain plant species. If a pathogen is capable of suppressing basal defense, plants may respond with another line of defense: the hypersensitive response (HR). The HR is characterized by deliberate plant cell suicide at the site of infection. Although drastic compared to basal resistance, the HR may limit pathogen access to water and nutrients by sacrificing a few cells in order to save the rest of the plant. The HR is typically more pathogen-specific than basal resistance and is often triggered when gene products in the plant cell recognize the presence of specific disease-causing effector molecules introduced into the host by the pathogen. Bacteria, fungi, viruses, and microscopic worms called nematodes are capable of inducing the HR in plants.

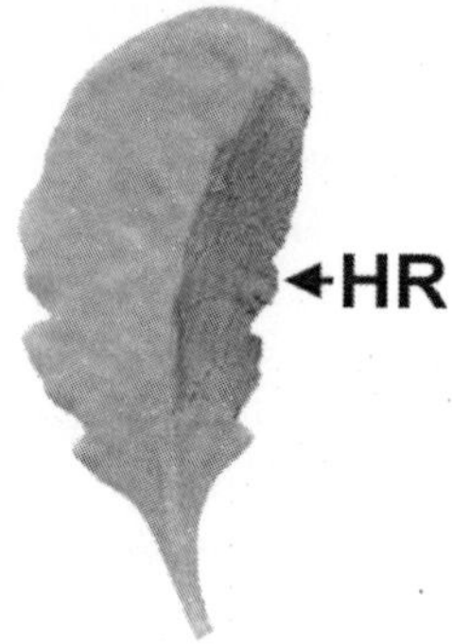

Figure: *HR lesion on an* Arabidopsis *leaf.*

Once the hypersensitive response has been triggered, plant tissues may become highly resistant to a broad range of pathogens for an extended period of time. This phenomenon is called systemic acquired resistance (SAR) and represents a heightened state of readiness in which plant resources are mobilized in case of further attack. Researchers have learned to artificially trigger SAR by spraying plants with chemicals called plant activators. These substances are gaining favour in the agricultural community because they are much less toxic to humans and wildlife than fungicides or antibiotics, and their protective effects can last much longer.

In addition to the hypersensitive response, plants can defend themselves against viruses by a variety of mechanisms including a sophisticated genetic defense system called RNA silencing. Many viruses produce double-stranded RNA or DNA during replication in a host cell. Plants can recognize these foreign molecules and respond by digesting the genetic strands into useless fragments and halting the infection. Plants that are infected with viruses will often exhibit chlorosis and mottling, but disease symptoms may eventually disappear if RNA silencing is successful, a process called recovery. In addition, the plant may retain a template of the digested genetic strand that can be used to quickly respond to future attack by similar viruses, a process analogous to the memory of vertebrate immune systems.

Detection of Insect Herbivores

Mechanical damage caused by insects is not generally considered "true" plant disease although plants have developed surveillance systems designed to recognize insect pests and respond with specific defense mechanisms. Plants can distinguish between general wounding and insect feeding by the presence of elicitors contained in the saliva of chewing insects. In response, plants may release volatile organic compounds (VOCs), including monoterpenoids, sesquiterpenoids, and homoterpenoids. These chemicals may repel harmful insects or attract beneficial predators that prey on the destructive pests.

For example, wheat seedlings infested with aphids may produce VOCs that repel other aphids. Lima beans and apple trees emit chemicals that attract predatory mites when damaged by spider mites, and cotton plants produce volatiles that attract predatory wasps when damaged by moth larvae. Feeding on one part of the plant can induce systemic production of these chemicals in undamaged plant tissues, and once released, these chemicals can act as signals to neighbouring plants to begin producing similar compounds. Production of these chemicals exacts a high metabolic cost on the host plant, so many of these compounds are not produced in large quantities until after insects have begun to feed.

Structural Defenses

The Plant Cell

All plant tissues contain pre-formed structural barriers that help limit pathogen attachment, invasion and infection. The cell wall is a major line of defense against fungal and bacterial pathogens. It provides an excellent structural barrier that also incorporates a wide variety of chemical defenses that can be rapidly activated when the cell detects the presence of potential pathogens. All plant cells have a primary cell wall, which provides structural support and is essential for turgor pressure, and many also form a secondary cell wall that develops inside of the primary cell wall after the cell stops growing.

The primary cell wall consists mostly of cellulose, a complex polysaccharide consisting of thousands of glucose monomers linked together to form long polymer chains. These chains are bundled into fibres called microfibrils, which give strength and flexibility to the wall. The cell wall may also contain two groups of branched polysaccharides: cross-linking glycans and pectins. Cross-linking glycans include hemicellulose fibres that give the wall strength via cross-linkages with cellulose. Pectins form hydrated gels that help "cement" neighbouring cells together and regulate the water content of the wall. Soft-rot pathogens often target pectins for digestion using specialized enzymes that cause cells to break apart: these organisms are extremely common, and anyone who has seen fruits or vegetables become brown and "mushy" have seen these pathogens in action.

Many cell walls also contain lignin, a heterogeneous polymer composed of phenolic compounds that gives the cell rigidity. Lignin is the primary component of wood, and cell walls that become "lignified" are highly impermeable to pathogens and difficult for small insects to chew. Cutin, suberin, and waxes are fatty substances that may be deposited in either primary or secondary cell walls (or both) and outer protective tissues of the plant body, including bark.

Cell walls contain proteins and enzymes that actively work to reshape the wall during cell growth yet thicken and strengthen the wall during induced defense. When a plant cell detects the presence of a potential pathogen, enzymes catalyze an oxidative burst that produces highly reactive oxygen molecules capable of damaging the cells of invading organisms. Reactive oxygen molecules also help strengthen the cell wall by catalyzing cross-linkages between cell wall polymers, and they serve as a signal to neighbouring cells that an attack is underway. Plant cells also respond to microbial attack by rapidly synthesizing and depositing callose between the cell wall and cell membrane adjacent to the invading pathogen. Callose deposits, called papillae, are polysaccharide polymers that impede cellular penetration at the site of infection, and these are often produced as part of the induced basal defense response.

Some plant cells are highly specialized for plant defense. Idioblasts ("crazy cells") help protect plants against herbivory because they contain toxic chemicals or sharp crystals that tear the mouthparts of insects and mammals as they feed. There are many classes of idioblasts including pigmented cells, sclereids, crystalliferous cells, and silica cells. Pigmented cells often contain bitter-tasting tannins that make plant parts undesirable as a food source. Young red wines often contain high levels of tannins that give wine a sharp, biting taste. Sclereids are irregularly-shaped cells with thick secondary walls that are difficult to chew: the rough texture of pear fruit (*Pyrus* spp.) is caused by thousands of sclereid stone cells that can abrasively wear down the teeth of feeding animals. Stinging nettles (*Urtica dioica*) produce stinging cells shaped like hypodermic needles that break off when disturbed and inject highly irritating toxins into herbivore tissues. Some stinging cells contain prostaglandins, hormones that amplify pain receptors in vertebrate animals and increase the sensation of pain. Crystalliferous cells contain crystals of calcium oxalate that may tear herbivore mouthparts when chewed and can be toxic if ingested. Members of the genera *Philodendron* and *Dieffenbachia* are very common tropical house plants that contain large amounts of these cells. Humans and pets who chew the leaves of these plants may experience a burning sensation in the mouth and throat that is often accompanied by swelling, choking, and an inability to speak. For these reasons, species of Dieffenbachia are commonly called dumb cane. Grasses and sedges contain rows of silica cells in their epidermal layers which give strength and rigidity to the growing leaf blades and deter feeding by chewing insects.

Plant Tissues and Specialized Appendages

The epidermis constitutes the outermost protective tissue system of leaves, floral parts, fruits, seeds, stems, and roots of plants until they undergo considerable secondary growth. It is the first line of defense against invading pathogens and consists of both specialized and unspecialized cells. The epidermal cells of aerial plant parts are often covered in a waxy cuticle that not only prevents water loss from the plant, but also prevents microbial pathogens from coming into direct contact with epidermal cells and thereby limits infection. The cuticle can be relatively thin (aquatic plants) or extremely thick (cacti). The hydrophobic nature of the cuticle also prevents water from collecting on the leaf surface, an important defense against many fungal pathogens that require standing water on the leaf surface for spore germination. However, some fungal pathogens including *Fusarium solani* produce cutinases that degrade the cuticle and allow the fungi to penetrate the epidermis.

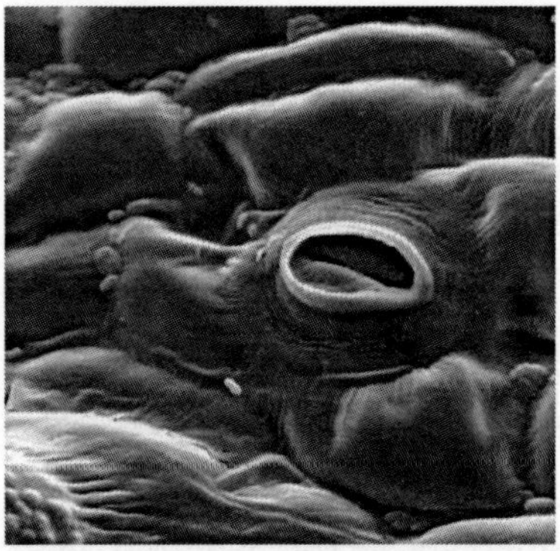

Figure: *Stomate on the surface of a tomato leaf. Note the small rod-shaped bacteria.*

Interspersed among the many unspecialized cells of the epidermis are guard cells which regulate gas exchange through small openings called stomata. These pores allow carbon dioxide to enter the leaf for use in photosynthesis while restricting excessive water loss from the plant. Stomatal pore size is highly regulated by plants, and guard cells can participate in defense by closing in response to the presence of MAMPs.

Trichomes ("leaf hairs") are specialized epidermal cells found on aerial plant parts that may provide both physical and chemical protection against insect pests. The velvety appearance of dusty miller (*Senecio cineraria*) is caused by thousands of tiny trichomes covering the plant's surface. Trichomes on the surface of soybeans (*Glycine max*) prevent insect eggs from reaching the epidermis and the larvae starve after hatching. The hook-shape of snap bean (*Phaseolis vulgaris*) trichomes impale caterpillars as they move across the leaf surface, and glandular trichomes in potato and tomato secrete oils that repel aphids. In woody plants, the periderm replaces the epidermis on stems and roots. Outer bark (phellem) is an excellent example of a preformed structural barrier that contains high amounts of water-resistant suberin and prevents many pathogens and insects from reaching the living cells underneath.

Thorns are modified branches that protect plants from grazing vertebrates, and include the honey locust tree (*Gleditsia triacanthos*). Many cacti produce thorn-like structures that are actually modified leaves or parts of leaves (e.g., stipules) called spines which serve similar purposes, such as in the barrel cactus (*Ferocactus* spp.). Botanically speaking, the "thorns" on the stem of rose plants (*Rosa* spp.) are neither true thorns nor spines: they are actually outgrowths of the epidermis called prickles.

Chemical Defenses

Plant chemicals can be divided into two major categories: primary metabolites and secondary metabolites. Primary metabolites are substances produced by all plant cells that are directly involved in growth, development, or reproduction. Examples include sugars, proteins, amino acids, and nucleic acids. Secondary metabolites are not directly involved in growth or reproduction but they are often involved with plant defense. These compounds usually belong to one of three large chemical classes: terpenoids, phenolics, and alkaloids.

Terpenoids (terpenes) occur in all plants and represent the largest class of secondary metabolites with over 22,000 compounds described. The simplest terpenoid is the hydrocarbon isoprene (C_5H_8), a volatile gas emitted during photosynthesis in large quantities by leaves that may protect cell membranes from damage caused by high temperature or light. Terpenoids are classified by the number of isoprene units used to construct them. For example, monoterpenoids consist of two isoprene units, sesquiterpenoids (three units), diterpenoids (four units), and triterpenoids (six units).

Monoterpenoids and sesquiterpenoids are the primary components of essential oils, which are highly volatile compounds that contribute to the fragrance (essence) of plants that produce them. Essential oils often function as insect toxins and many protect against fungal or bacterial attack. Mint plants (*Mentha* spp.) produce large quantities of the monoterpenoids menthol and menthone which are produced and stored in glandular trichomes on the epidermis. Pyrethrins are monoterpenoid esters produced by chrysanthemum plants that act as insect neurotoxins. Many commercially available insecticides are actually synthetic analogues of pyrethrins, called pyrethroids, including the insecticides permethrin and cypermethrin. Pine tree resin contains large quantities of the monoterpenoids alpha- and beta-pinene, which are potent insect repellents; these compounds give the organic solvent turpentine its characteristic sharp odor.

Monoterpenoids are not just used as insecticides. Many spices, seasonings, condiments, and perfumes are made using essential oils that function as insect toxins in plants but are relatively harmless to humans. Examples include peppermint and spearmint (*Mentha* spp.), basil (*Ocimum* spp.), oregano (*Origanum* spp.), rosemary (*Rosmarinus* spp.), sage (*Salvia* spp.), savory (*Satureja* spp.), thyme (*Thymus* spp.), black pepper (*Piper* spp.), cinnamon (*Cinnamomum* spp.), and bay leaf (*Laurus* spp.).

Diterpenoids include gossypol, a terpenoid produced by cotton (*Gossypium hirsutum*) that has strong antifungal and antibacterial properties. Triterpenoids are similar in molecular structure to plant and animal sterols and steroid hormones. Phytoectysones are mimics of insect molting hormones. When produced by plants such as spinach (*Spinacia oleracea*), they disrupt larval development and increase insect mortality. The fresh scent of lemon and orange peels is the result of a class of triterpenoids called limonoids. Azadirachtin is a very powerful limonoid isolated from neem trees (*Azadirachta indica*): some insects are repelled by concentrations as low as a few parts per million. Citronella is an essential oil isolated from lemon grass (*Cymbopogon citratus*); it contains high limonoid levels and has become a popular insect repellent in the United States due to its low toxicity in humans and biodegradable properties.

Insects are not the only herbivores that feed on plants. Triterpenoids such as cardiac glycosides are highly toxic to vertebrate herbivores, including humans, and can cause heart attacks if ingested in high quantities. Foxglove (*Digitalis purpurea*) is the principal source of the cardiac glycosides digitoxin and digoxin, which are used medicinally in small quantities to treat heart disease in people. Some herbivores have overcome the dangerous effects of cardiac glycosides and actually use these toxins for their own benefit. Monarch butterfly caterpillars feed almost exclusively upon milkweed (*Asclepias* spp.) which contains high amounts of these toxins in the milky latex of their sap. The caterpillars store these toxins safely within their bodies, and when the caterpillars become adult butterflies, they are highly poisonous to most predatory birds that eat them.

Figure: *Digitalis purpurea*

Saponins are glycosylated triterpenoids (triterpenoids with attached sugar groups) that are present in the cell membranes of many plant species. These substances have detergent (soap-like) properties and disrupt the cell membranes of invading fungal pathogens. The wheat pathogen *Gaeumannomyces graminis* is unable to infect oats that contain avenacins, a class of triterpenoid saponins. However, some fungal pathogens have developed counter-measures to these plant defenses: Botrytis cinerea, Fusarium oxysporum, and *Septoria lycopersici* are all capable of degrading saponins and causing disease in susceptible saponin-producing plants.

Phenolics

Phenolics are another large class of secondary metabolites produced by plants to defend themselves against pathogens. They are produced primarily via the shikimic acid and malonic acid pathways in plants, and include a wide variety of defense-related compounds including flavonoids, anthocyanins, phytoalexins, tannins, lignin, and furanocoumarins. Flavonoids are one of the largest classes of phenolics. Anthocyanins are colourful water-soluble flavonoids pigments produced by plants to protect foliage from the damaging effects of ultraviolet radiation. Anthocyanins are responsible for the showy colours of many plants and are present in high concentrations in flowers, fruits, and the leaves of deciduous plants in fall. Phytoalexins are isoflavonoids with antibiotic and antifungal properties that are produced in response to pathogen attack. These toxic molecules disrupt pathogen metabolism or cellular structure but are often pathogen specific in their toxicity. Examples include medicarpin produced by alfalfa (*Medicago sativa*), rishitin produced by both tomatoes and potatoes (the *Solanaceae* family), and camalexin, produced by *Arabidopsis thaliana*.

Tannins are water-soluble flavonoid polymers produced by plants and stored in vacuoles. Tannins are toxic to insects because they bind to salivary proteins and digestive enzymes including trypsin and chymotrypsin resulting in protein inactivation. Insect herbivores that ingest high amounts of tannins fail to gain weight and may eventually die. The sharp taste of red wine is caused by grape tannins binding to salivary proteins in the mouth which results in protein coagulation.

Lignin is a highly branched heterogeneous polymer found principally in the secondary cell walls of plants, although primary walls can also become lignified. It consists of hundreds or thousands of phenolic monomers and is a primary component of wood. Because it is insoluble, rigid, and virtually indigestible, lignin provides an excellent physical barrier against pathogen attack.

Furanocoumarins are phenolic compounds produced by a wide variety of plants in response to pathogen or herbivore attack. They are activated by ultraviolet light and can be highly toxic to certain vertebrate and invertebrate herbivores due to their integration into DNA, which contributes to rapid cell death. In fact, grapefruit juice contains small quantities of furanocoumarins, which greatly increase the absorption of certain drugs into the bloodstream from the intestines. Some medicines carry warning labels cautioning people to avoid drinking grapefruit juice while taking the drugs in order to avoid an accidental overdose.

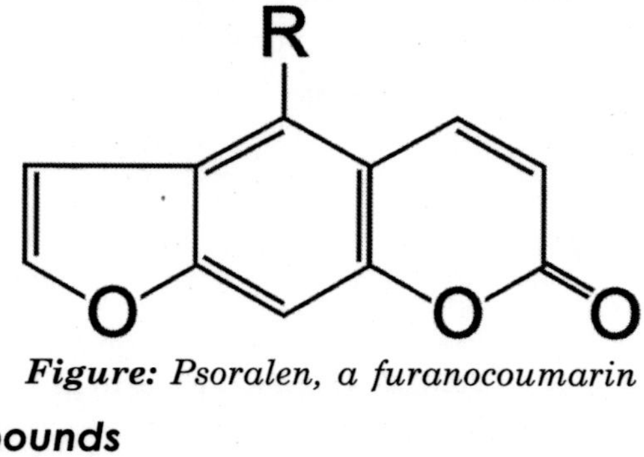

***Figure:** Psoralen, a furanocoumarin*

Nitrogen Compounds

Alkaloids are a large class of bitter-tasting nitrogenous compounds that are found in many vascular plants and include caffeine, cocaine, morphine, and nicotine. They are derived from the amino acids aspartate, lysine, tyrosine, and tryptophan, and many of these substances have powerful effects on animal physiology. Caffeine is an alkaloid found in plants such as coffee (*Coffea arabica*), tea (*Camellia sinensis*), and cocoa (*Theobroma cacao*). It is toxic to both insects and fungi. In fact, high levels of caffeine produced by coffee seedlings can even inhibit the germination of other seeds in the vicinity of the growing plants, a phenomenon called allelopathy. Allelopathy allows one plant species to "defend" itself against other plants that may compete for growing space and nutrient resources.

Members of the nightshade family (*Solanaceae*) produce many important alkaloid compounds. Nicotine is an alkaloid that is produced in the roots of tobacco plants (*Nicotiana tabacum*) and transported to leaves where it is stored in vacuoles. It is released when herbivores graze on the leaves and break open the vacuoles. Atropine is a neurotoxin and cardiac stimulant produced by the deadly nightshade plant (*Atropa belladonna*). Although it is toxic in large quantities, it has been used medicinally by humans in small amounts as a pupil dilator and antidote for some nerve gas poisonings. Capsaicin and related capsaicinoids produced by members of the genus *Capsicum* are the active components of chili peppers and produce their characteristic burning sensation in hot, spicy foods.

Cyanogenic glycosides are a particularly toxic class of nitrogenous compounds that break down to produce hydrogen cyanide (HCN), a lethal chemical that halts cellular respiration in aerobic organisms. Plants that produce cyanogenic glycosides also produce enzymes that convert these compounds into hydrogen cyanide, including glycosidases and hydroxynitrile lyases, but they are stored in separate compartments or tissues within the plant; when herbivores feed on these tissues, the enzymes and substrates mix and produce lethal hydrogen cyanide. Glucosinolates, also known as mustard oil glycosides, are sulfur-containing compounds synthesized by members of the mustard family (*Brassicaceae*) and produce cyanide gas when broken down by enzymes called thioglucosidases.

Proteins and Enzymes

Many plants and seeds contain proteins that specifically inhibit pathogen and pest enzymes by forming complexes that block active sites or alter enzyme conformations, ultimately reducing enzyme function. These proteins are generally small and rich in the amino acid cysteine. They include defensins, amylase inhibitors, lectins, and proteinase inhibitors. Unlike simple chemicals such as terpenoids, phenolics, and alkaloids, proteins require a great deal of plant resources and energy to produce; consequently, many defensive proteins are only made in significant quantities after a pathogen or pest has attacked the plant. Once activated, however, defensive proteins and enzymes effectively inhibit fungi, bacteria, nematodes, and insect herbivores.

Defensins are small cysteine-rich proteins that display broad antimicrobial activity and were first isolated from the endosperm of barley (*Hordeum vulgare*) and wheat (*Triticum aestivum*). They are widely distributed and may be present in most plants. Defensins are best characterized in seeds, but can be found in virtually all types of plant tissues including leaves, pods, tubers, fruit, roots, bark, and floral tissues. They exhibit a wide range of biological activities that serve to inhibit the growth of many fungi and bacteria. Some defensins also inhibit digestive proteins in herbivores. The precise mechanisms employed by plant defensins to inhibit fungi and bacteria are still being characterized, but they appear to act upon molecular targets in the plasma membrane of pathogens. These defensins may inhibit pre-existing ion channels or form new membrane pores that disrupt cellular ion balance.

Figure: *Wheat (*Triticum aestivum*)*

Figure: *Castor beans (Ricinus communis)*

Digestive enzyme inhibitors are proteins that block the normal digestion and absorption of nutrients by vertebrate and invertebrate herbivores. Alpha-amylase inhibitors are proteins commonly found in legumes that bind to amylase enzymes and inhibit starch digestion. Lectins are non-enzymatic proteins and glycoproteins that bind to carbohydrates and exhibit a wide range of functions including disruption of digestion in insects and agglutination of blood cells in vertebrates.

Ricin is a powerful toxin produced in castor beans (*Ricinus communis*). It combines a lectin molecule with an N-glycoside hydrolase that enters animal cells and inhibits protein synthesis. Ricin is a highly potent toxin, having an average lethal dose of only 0.2 milligrams in humans.

Protease inhibitors are typically produced in response to herbivore attack and inhibit digestive enzymes including trypsin and chymotrypsin. They occur widely in nature but have been well studied in legumes, solanaceous plants, and grasses. Herbivore feeding often triggers a series of molecular signaling events that induce systemic production of these compounds in distal tissues that contribute to the protection of undamaged plant parts from subsequent attacks by a wide range of herbivore pests.

Hydrolytic enzymes are produced by some plants in response to pathogens and often accumulate in extracellular spaces where they degrade the cell walls of pathogenic fungi. Chitinases are enzymes that catalyze the degradation of chitin, a polymer with a backbone similar to cellulose that is present in the cell walls of true fungi. Glucanases are enzymes that catalyze the degradation of glycosidic linkages in glucans, a class of polymers similar to cellulose that is present in the cell walls of many oomycetes (water molds). In vitro analysis has verified the anti-fungal properties of these compounds, and transgenic plants expressing high levels of these enzymes exhibit increased resistance to a wide range of both foliar and root pathogens. Lysozymes are hydrolytic enzymes that are capable of degrading bacterial cell walls.

Bonsai

Bonsai is a Japanese art form using miniature trees grown in containers. Similar practices exist in other cultures, including the Chinese tradition of *penjing* from which the art originated, and the miniature living landscapes of Vietnamese *hòn non b*Ω. The Japanese tradition dates back over a thousand years, and has its own aesthetics and terminology.

"Bonsai" is a Japanese pronunciation of the earlier Chinese term *penzai*. A "bon" is a tray-like pot typically used in bonsai culture. The word *bonsai* is often used in English as an umbrella term for all miniature trees in containers or pots. This article focuses on bonsai as defined in the Japanese tradition.

The purposes of bonsai are primarily contemplation (for the viewer) and the pleasant exercise of effort and ingenuity (for the grower). By contrast with other plant cultivation practices, bonsai is not intended for production of food, for medicine, or for creating yard-size or park-size gardens or landscapes. Instead, bonsai practice focuses on long-term cultivation and shaping of one or more small trees growing in a container.

A bonsai is created beginning with a specimen of source material. This may be a cutting, seedling, or small tree of a species suitable for bonsai development. Bonsai can be created from nearly any perennial woody-stemmed tree or shrub species that produces true branches and can be cultivated to remain small through pot confinement with crown and root pruning. Some species are popular as bonsai material because they have characteristics, such as small leaves or needles, that make them appropriate for the compact visual scope of bonsai.

The source specimen is shaped to be relatively small and to meet the aesthetic standards of bonsai. When the candidate bonsai nears its planned final size it is planted in a display pot, usually one designed for bonsai display in one of a few accepted shapes and proportions. From that point forward, its growth is restricted by the pot environment. Throughout the year, the bonsai is shaped to limit growth, redistribute foliar vigor to areas requiring further development, and meet the artist's detailed design.

The practice of bonsai is sometimes confused with dwarfing, but dwarfing generally refers to research, discovery, or creation of plant cultivars that are permanent, genetic miniatures of existing species. Bonsai does not require genetically dwarfed trees, but rather depends on growing small trees from regular stock and seeds. Bonsai uses cultivation techniques like pruning, root reduction, potting, defoliation, and grafting to produce small trees that mimic the shape and style of mature, full-size trees.

History of Bonsai

Early versions

The Japanese art of bonsai originally derived from the Chinese practice of penjing. From the 6th century onwards, Imperial embassy personnel and Buddhist students from Japan had been visiting and returning from mainland China, bringing back souvenirs that included container plantings. At least 17 diplomatic missions were specifically sent from Japan to the Tang court between the years 603 and 839.

Japan's historical Shôsôin, which houses 7th, 8th, and 9th-century artifacts including material from Japan's Tempyô period, contains an elaborate miniature tree display dating from this time. This artifact is composed of a wooden tray serving as a base, carved wooden mountain models, and sand portraying a riverine sandbar. Small tree sculptures in silver metal are meant to be placed in the sand, to produce a table-top depiction of a treed landscape. While this display is closer to the Japanese bonkei display than to a living bonsai, it does reflect the period's interest in miniature landscapes.

From about the year 970 comes the first lengthy work of fiction in Japanese, *Utsubo Monogatari* (*The Tale of the Hollow Tree*), which includes this passage: "A tree that is left growing in its natural state is a crude thing. It is only when it is kept close to human beings who fashion it with loving care that its shape and style acquire the ability to move one." The idea, therefore, was already established by this time that natural beauty becomes true beauty only when modified in accordance with a human ideal.

In the medieval period, recognizable bonsai began to appear in handscroll paintings like the *Ippen shonin eden* (1299). *Saigyo Monogatari Emaki* was the earliest known scroll to depict dwarfed potted trees in Japan. It dates from the year 1195, in the Kamakura period. Wooden tray and dish-like pots with dwarf landscapes on modern-looking wooden shelf/benches are shown in the 1309 *Kasuga-gongen-genki* scroll. These novelties show off the owner's wealth and were probably exotics imported from China.

Chinese Chan Buddhist monks also came over to teach at Japan's monasteries, and one of the monks' activities was to introduce political leaders of the day to the various arts of miniature landscapes as ideal accomplishments for men of taste and learning.

The c.1300 rhymed prose essay, *Bonseki no Fu* (*Tribute to Bonseki*) written by celebrated priest and master of Chinese poetry, Kokan Shiren (1278–1346), outlined the aesthetic principles for what would be termed bonsai, bonseki and garden architecture itself. At first, the Japanese used miniaturized trees grown in containers to decorate their homes and gardens.

Criticism of the interest in curiously twisted specimens of potted plants shows up in one chapter of the 243-chapter compilation *Tsurezuregusa* (c.1331). This work would become a sacred teaching handed down from master to student, through a limited chain of poets (some famous), until it was at last widely published in the early 17th century. Before then, the criticism had only a modest influence on dwarf potted tree culture.

In 1351, dwarf trees displayed on short poles were portrayed in the *Boki Ekotoba* scroll. Several other scrolls and paintings also included depictions of these kinds of trees. Potted landscape arrangements made during the next hundred years or so included figurines after the Chinese fashion in order to add scale and theme. These miniatures would eventually be considered garnishes decidedly to be excluded by Japanese artists who were simplifying their creations in the spirit of Zen Buddhism.

Hachi-no-ki

***Figure:** Japanese white pine from the National Bonsai & Penjing Museum at the United States National Arboretum.*

Around the 14th century, the term for dwarf potted trees was "the bowl's tree" (b'n0(g, *hachi-no-ki*). This denoted the use of a fairly deep pot, as opposed to the shallow pot denoted by the term *bonsai.*

Hachi-No-Ki (*The Potted Trees*) is also the title of a Noh play by Zeami Motokiyo (1363–1444), based on a story from c. 1383. It tells of an impoverished samurai who sacrifices his three last dwarf potted trees as firewood to provide warmth for a travelling monk on a winter night. The monk is an official in disguise who later rewards the samurai by giving him three lands whose names include the names of the three types of trees the samurai burnt: *ume* (plum), *matsu* (pine), and *sakura* (cherry). In later centuries, woodblock prints by several artists would depict this popular drama. There was even a fabric design of the same name.

Stories referring to bonsai began to appear more frequently by the 17th century. *Shogun* Tokugawa Iemitsu (r. 1623-1651) was a *hachi-no-ki* enthusiast. A story tells of Okubo Hikozemon (1560–1639), councilor to the shogun, who threw one of Iemitsu's favourite trees away in the garden—in sight of the shogun—in order to dissuade him from spending so much time and attention on these trees.

In spite of the servant's efforts, Iemitsu never gave up his beloved art form. Another story from this time tells of a samurai's gardener who killed himself when his master insulted a hachi-no-ki of which the artisan was especially proud.

Bonsai dating to the 17th century have survived to the present. One of the oldest-known living bonsai trees, considered one of the National Treasures of Japan, is in the Tokyo Imperial Palace collection. A five-needle pine (*Pinus pentaphylla* var. *negishi*) known as Sandai-Shogun-No Matsu is documented as having been cared for by Tokugawa Iemitsu. The tree is thought to be at least 500 years old and was first trained as a bonsai by, at latest, the year 1610. The earliest known report by a Westerner of a Japanese dwarf potted tree was made in 1692 by George Meister.

Chinese bonsai containers exported to Japan during the 17th and 18th centuries would become referred to as *Kowatari* ("old crossing"). These were made between 1465 and about 1800. Many came from Yixing in Jiangsu province—unglazed and usually purplish-brown—and some others from around Canton, in particular, during the Ming dynasty. Miniature potted trees were called *hachi-ue* in a 1681 horticulture book. This book also stated that everyone at the time grew azaleas, even if the poorest people had to use an abalone shell as a container. Torii Kiyoharu's use of woodblock printing in Japan depicted the dwarf potted trees from horticultural expert Itô Ihei's nursery.

By the end of the 18th century, bonsai cultivation was quite widespread and had begun to interest the public. In the Tenmei era (1781–88), an exhibit of traditional dwarf potted pines began to be held every year in Kyoto. Connoisseurs from five provinces and neighbouring areas would bring one or two plants each to the show in order to submit them to visitors for ranking.

Classical Period of Bonsai

In Itami, Hyogo (near Osaka), a group of scholars of Chinese arts gathered in the early 19th century to discuss recent styles in the art of miniature trees. Their version of these, which had been previously called "Bunjin Ueki," "Bunjin Hachiue," or other terms, were renamed "bonsai" (the Japanese pronunciation of the Chinese term penzai). This term had the connotation of a shallower container in which the Japanese could now more successfully style small trees. The term "bonsai," however, would not become regularly used in describing their dwarf potted trees for nearly a century. Many others terms and compositions adopted by this group were derived from *Kai-shi-en Gaden,* the Japanese version of *Jieziyuan Huazhuan (Manual of the Mustard Seed Garden).*

In 1829, a significant book that first established classical bonsai art, *Somoku Kinyo Shu* (*A Colourful Collection of Trees and Plants/ Collection of tree leaves*), was published. It includes the basic criteria for the ideal form of the classical pine bonsai, in detail and with illustrations. That same year, small *tako-tsuki* (octopus-styled) trees with long, wavy-branches began to be offered by a grower in Asakusa Park, a north-eastern Edo suburb. Within 20 years that neighbourhood became crowded with nurseries selling bonsai. The three-volume *Kinsei-Jufu,* possibly the first book of bonsai, tools, and pots, dates from 1833.

Numerous artists of the 19th century depicted dwarf potted trees in woodblock prints, including Yoshishige (who pictured each of the fifty-three classic stations of the Tokaido (road) as miniature landscape) and Kunisada (who included mostly *hachi-no-ki* in some four dozen prints). The earliest known photograph from Japan depicting a dwarf potted tree dates from c. 1861 by Pierre Rossier.

On October 13, 1868, the Meiji Emperor moved to his new capital in Tokyo. Bonsai were displayed both inside and outside Meiji Palace, where they have since remained important in affairs of the Palace. Bonsai placed in the grand setting of the Imperial Palace had to be "Giant Bonsai," large enough to fill the grand space. The Meiji Emperor encouraged interest in bonsai. Government officials who did not appreciate bonsai fell out of favour. Soon all members of the ministry had bonsai whether they liked the tradition or not. Prince Itoh was an exception: Any bonsai that the emperor gave him were then passed to Kijoji Itoh. Kijoji Itoh was a statesman of great influence behind the scenes, and a noted bonsai collector who conducted research and experiments on these bonsai.

Bonsai shaping aesthetics and techniques were becoming more sophisticated. By the late 1860s, thick combed and wetted hemp fibres were used to roughly shape the trunk and branches of miniature trees by pulling and tying them. The process was tedious and bothersome, and the final product was unsightly. Tips of branches would only be opened flat. Long, wavy-branched *tako* (octopus)-style trees were mass-produced and designed in the [renamed capital] Tokyo for the increasing foreign trade, while the more subtle and delicate bunjin-style trees designed in Kyoto and Osaka were for use in Japan. Tokyo preferred big trunks out of proportion and did not approve of Kyoto's finely designed slender trunks. (This cultural rivalry would continue for a century.)

Pots exported from China between 1816 and 1911 (especially the late 19th century) were called *Nakawatari* (*middle-crossing*) or *Chuwatari,* shallow rectangular or oval stoneware with carved feet and drainage holes. Unglazed pots of this type were used at ancestral shrines and treasured by the Chinese. After the mid-century, certain Japanese antiquities dealers imported them and instant popular approval for this type of container for bonsai created a huge demand. As a consequence, orders came from Japan to Yixing pottery centres specifically to make bonsai pots.

Through the later 19th century, Japanese participation in various international exhibitions introduced many in the U.S. and Europe to dwarf potted trees. Specimens from the displays went into Western hands following the closing of the fairs. Japanese immigrants to the U.S. West Coast and Hawaii Territory brought plants and cultivation experience with them. Export nurseries, the most notable one being the Yokohama Gardeners Association, provided increasingly good quality dwarf potted trees for Americans and Europeans—even if the buyers did not have enough information or experience to actually keep the trees alive long-term.

An Artistic Bonsai Concours was held in Tokyo in 1892 followed by the publication of a three-volume commemorative picture book. This demonstrated a new tendency to see bonsai as an independent art form. In 1903, the Tokyo association Jurakukai held showings of bonsai and ikebana at two Japanese-style restaurants. Three years later, *Bonsai Gaho* (1906 to c.1913), became the first monthly magazine on the subject. It was followed by *Toyo Engei* and *Hana* in 1907, and *Bonsai* in 1921.

By 1907, "on the outskirts of Tokio [dwarf] tree artists have formed a little colony of from twenty to thirty houses, and from this centre their work finds its way to all parts of the world." "Its secrets are handed down from father to son in a few families, and are guarded with scrupulous care."

Count Okuma (1838–1922) maintained a famed collection of dwarf pines and dwarf plum trees.

In 1910, shaping with wire was described in the Sanyu-en Bonsai-Dan (History of Bonsai in the Sanyu nursery). Zinc-galvanized steel wire was initially used. Expensive copper wire was only used for trees that had real potential. Between 1911 and about 1940, mass-produced containers were exported from Yixing, China, and made to the specifications of Japanese dealers. These were called *Shinto* (*new crossing or arrival*) or *Shin-watare.* These were made for increasing numbers of enthusiasts. Some containers, including primitive style ones, were also being made in Formosa.

By 1914, "at the N.E. corner of Shiba Park is a permanent bazaar (the first of its kind established in Tokyo) where hosts of native-made gimcracks can be bought at fixed prices. The exhibits of potted plants and dwarf trees held here from time to time attract lovers of such things." Also this year, the first national annual bonsai show was held (through 1933) in Tokyo's Hibiya Park. During this period, the tokonoma in formal rooms and tea rooms became the main place for bonsai display. The shaped trees now shared space with other items such as scrolls, incense burners, Buddhist statues and tea ceremony implements.

The first issue of *Bonsai* magazine was published in 1921 by Norio Kobayashi (1889–1972). This influential periodical would run for 518 consecutive issues. Copper wire was being extensively used by this time. Major changes to a tree's shape could now be accomplished with wiring. Trees could be precisely and aesthetically wired, and then sold immediately. A greater number of both collected and nursery trees could now be trained for bonsai. The number of hobbyists increased due to the increased ability to style with wire, but there was also an increase in damaged or scarred trees.

The 1923 Great Kanto Earthquake and resulting fire devastated Tokyo, and gutted the downtown area where many bonsai were grown. And so, two years later, a group of thirty families of downtown Tokyo professional growers established the Ômiya Bonsai Village, northeast of the capital. The first great annual public exhibition of trees was held at the Asahi Newspaper Hall in Tokyo in 1927. The first of the very prestigious Kokufu-ten exhibitions were held in Tokyo's Ueno Park, beginning in 1934. By the following year, tokonoma display principles allowed for bonsai to be shown for the tree's individual beauty, not just for its spiritual or symbolic significance.

Toolsmith Masakuni I (1880–1950) helped design and produce the first steel tools specifically made for the developing requirements of bonsai styling.

By 1940, there were about 300 bonsai dealers in Tokyo, some 150 species of trees were being cultivated, and thousands of specimens annually were shipped to Europe and America. The first major book on the subject in English was published in the Japanese capital: *Dwarf Trees (Bonsai)* by Shinobu Nozaki (1895–1968). The first bonsai nurseries and clubs in the Americas were started by first and second-generation Japanese immigrants.

Caretaker of the Imperial bonsai collection, Kyuzo Murata (1902–1991), was one of very few persons allowed to take care of bonsai during the Pacific War. He gathered together and preserved many trees from the other Omiya growers and would water them under the protection of night. Throughout 1945, many old trees were the smallest casualties of the spring and summer napalm bombing of Tokyo (esp. March 9/10) and sixty-six other cities. Gardeners protected the Imperial collection trees from fire by pouring water over them after the Palace caught fire when neighbouring areas were bombed on May 25/26. Following the surrender of Japan, there began the post-war re-evaluation and reviving of damaged collections of trees—including the Imperial—which would continue for over a decade as Japan was rebuilt. Many of the Omiya growers did not continue their vocation.

During the Allied Occupation of Japan (through 1952) U.S. officers and their wives could take courses in bonsai, bonkei, ikebana, and other traditional arts and crafts as arranged by General MacArthur's headquarters. Many of the older and limited varieties of trees were no longer available, and the bonsai considered in fashion changed partly because of this shortage. Copper wire now largely replaced ordinary iron wire for shaping the better trees, but the latter still would be used for mass-produced commercial bonsai.

Modern Bonsai

Following World War II, a number of trends made the Japanese tradition of bonsai increasingly accessible to Western and world audiences. One key trend was the increase in the number, scope, and prominence of bonsai exhibitions. For example, the Kokufu-ten bonsai displays reappeared in 1947 after a four-year cancellation and became annual affairs. These displays continue to this day, and are by invitation only for eight days in February. In October 1964, a great exhibition was held in Hibya Park by the private Kokufu Bonsai Association, reorganized into the Nippon Bonsai Association, to mark the Tokyo Olympics. A commemorative album titled *Gems of Bonsai and Suiseki* was published in Japanese and English. Other countries began presenting bonsai exhibitions as well, with recurring events now taking place in Taiwan and a number of other Asian countries, Australia, the United States, several European countries, and others. Currently, Japan continues to host regular exhibitions with the world's largest numbers of bonsai specimens and the highest recognized specimen quality.

Another key trend was the increase in books on bonsai and related arts, now being published for the first time in English and other languages for audiences outside Japan. In 1952, Yuji Yoshimura, son of a leader in the Japanese bonsai community, collaborated with German diplomat and author Alfred Koehn to give bonsai demonstrations. The first formal bonsai courses opened to the public and outsiders in Tokyo. Koehn had been an enthusiast before the war, and his 1937 book *Japanese Tray Landscapes* had been published in English in Peking. Yoshimura's 1957 book *The Art of Bonsai*, written in English with his student Giovanna M. Halford, addressed both cultivation and aesthetic aspects of bonsai growing and went on to be called the "classic Japanese bonsai bible for westerners" with over thirty printings.

The related art of saikei was introduced to English-speaking audiences in 1963 in Kawamoto and Kurihara's *Bonsai-Saikei.* This book described tray landscapes made with younger plant material than was traditionally used in bonsai, providing an alternative to the use of large, older plants, few of which had escaped war damage.

Other works in Japanese and English had been published by this time, and afterwards a tremendous number of books saw print. Translations and original volumes in over two dozen languages were published over the following decades. Once Japanese was no longer the sole language of bonsai, the number of clubs outside of Asia increased and interaction increased between members of all levels of experience.

A third trend was the increasing availability of expert bonsai training, at first only in Japan and then more widely. In 1967 the first group of Westerners studied at an Ômiya nursery. Returning to the U.S., these people established the American Bonsai Society. Other groups and individuals from outside Asia then visited and studied at the various Japanese nurseries, occasionally even apprenticing under the masters. These visitors brought back to their local clubs the latest techniques and styles, which were then further disseminated. Japanese teachers also travelled widely, bringing hands-on bonsai expertise to all six continents.

By the beginning of the 1970s, these trends were beginning to merge. A large display of bonsai and suiseki was held as part of Expo '70, and formal discussion was made of an international association of enthusiasts. Three monthly magazines were started this year: *Bonsai Sekai, Satsuki Kenkyu,* and *Shizen to Bonsai.* In 1975, the first Gafu-ten (Elegant-Style Exhibit) of *shohin* bonsai (13-25 cm (9.84 in) (5–10 in) tall) was held. So was the first Sakufu-ten (Creative Bonsai Exhibit), the only event in which professional bonsai growers exhibit traditional trees under their own names rather than under the name of the owner. It was organized by Hideo Kato (1918–2001) at Daimaru Department Store in Tokyo.

The First World Bonsai Convention was held in Osaka during the World Bonsai and Suiseki Exhibition in 1980. Nine years later, the first World Bonsai Convention was held in Omiya and the World Bonsai Friendship Federation (WBFF) was inaugurated. These conventions attracted several hundreds of participants from dozens of countries and have since been held every four years at different locations around the globe: 1993, Orlando, Florida; 1997, Seoul, Korea; 2001, Munich, Germany; 2005, Washington, D.C.; 2009, San Juan, Puerto Rico.

The final trend supporting world involvement in bonsai is the widening availability of specialized bonsai plant stock, soil components, tools, pots, and other accessory items. Bonsai nurseries in Japan advertise and ship specimen bonsai world-wide. Most countries have local nurseries providing plant stock as well, although finding specimen bonsai is more difficult outside Japan and bonsai enthusiasts will often start with local trees that have not been pre-shaped into candidate bonsai. Japanese bonsai soil components, such as Akadama clay, are available worldwide, and local suppliers also provide similar materials in many locations. Specialized bonsai tools are widely available from Japanese and Chinese sources. Potters around the globe provide material to hobbyists and specialists in many countries.

Bonsai has now definitively reached a world-wide audience. There are over twelve hundred books on bonsai and the related arts in at least twenty-six languages available in over ninety countries and territories. A few dozen magazines in over thirteen languages are in print. Several score of club newsletters are available on-line, and there are at least that many discussion forums and blogs. Educational videos and just the appearance of dwarf potted trees in films and on television reach a wide audience.

There are at least a hundred thousand enthusiasts in some fifteen hundred clubs and associations worldwide, as well as over five million unassociated hobbyists. Plant material from every location is being trained into bonsai and displayed at local, regional, national, and international conventions and exhibitions for enthusiasts and the general public.

Cultivation and Care

Bonsai cultivation and care requires techniques and tools that are specialized to support the growth and long-term maintenance of trees in small containers.

Material Sources

All bonsai start with a specimen of source material, a plant that the grower wishes to train into bonsai form. Bonsai practice is an unusual form of plant cultivation in that growth from seeds is rarely used to obtain source material. To display the characteristic aged appearance of a bonsai within a reasonable time, the source plant is often mature or at least partially grown when the bonsai creator begins work. Sources of bonsai material include:

- *Propagation* from a source tree through cuttings or layering.
- *Nursery stock* directly from a nursery, or from a garden centre or similar resale establishment.
- *Commercial bonsai growers*, which, in general, sell mature specimens that display bonsai aesthetic qualities already.
- *Collecting* suitable bonsai material in its original wild situation, successfully moving it, and replanting it in a container for development as bonsai. These trees are called yamadori and are often the most expensive and prized of all Bonsai.

Techniques

***Figure:** This juniper makes extensive use of both jin (deadwood branches) and shari (trunk deadwood).*

The practice of bonsai development incorporates a number of techniques either unique to bonsai or, if used in other forms of cultivation, applied in unusual ways that are particularly suitable to the bonsai domain. These techniques include:

- *Leaf trimming*, the selective removal of leaves (for most varieties of deciduous tree) or needles (for coniferous trees and some others) from a bonsai's trunk and branches.
- *Pruning* the trunk, branches, and roots of the candidate tree.
- *Wiring* branches and trunks allows the bonsai designer to create the desired general form and make detailed branch and leaf placements.
- *Clamping* using mechanical devices for shaping trunks and branches.
- *Grafting* new growing material (typically a bud, branch, or root) into a prepared area on the trunk or under the bark of the tree.

- *Defoliation*, which can provide short-term dwarfing of foliage for certain deciduous species.
- *Deadwood bonsai techniques* called *jin* and *shari* simulate age and maturity in a bonsai.

Care

Small trees grown in containers, like bonsai, require specialized care. Unlike houseplants and other subjects of container gardening, tree species in the wild, in general, grow roots up to several metres long and root structures encompassing several thousand litres of soil. In contrast, a typical bonsai container is under 25 centimetres in its largest dimension and 2 to 10 litres in volume. Branch and leaf (or needle) growth in trees is also of a larger scale in nature. Wild trees typically grow 5 metres or taller when mature, whereas the largest bonsai rarely exceed 1 metre and most specimens are significantly smaller. These size differences affect maturation, transpiration, nutrition, pest resistance, and many other aspects of tree biology. Maintaining the long-term health of a tree in a container requires some specialized care techniques:

- *Watering* must be regular and must relate to the bonsai species' requirement for dry, moist, or wet soil.
- *Repotting* must occur at intervals dictated by the vigour and age of each tree.
- *Tools* have been developed for the specialized requirements of maintaining bonsai.
- *Soil composition and fertilization* must be specialized to the needs of each bonsai tree, although bonsai soil is almost always a loose, fast-draining mix of components.
- *Location and overwintering* are species-dependent when the bonsai is kept outdoors as different species require different light conditions. It is important to note that few of the traditional bonsai species can survive inside a typical house, due to the usually dry indoor climate.

Aesthetics

Bonsai aesthetics are the aesthetic goals characterizing the Japanese tradition of growing an artistically shaped miniature tree in a container. Many Japanese cultural characteristics, in particular the influence of Zen Buddhism and the expression of Wabi-sabi, inform the bonsai tradition in Japan. Established art forms that share some aesthetic principles with bonsai include penjing and saikei. A number of other cultures around the globe have adopted the Japanese aesthetic approach to bonsai, and, while some variations have begun to appear, most hew closely to the rules and design philosophies of the Japanese tradition.

Over centuries of practice, the Japanese bonsai aesthetic has encoded some important techniques and design guidelines. Like the aesthetic rules that govern, for example, Western common practice period music, bonsai's guidelines help practitioners work within an established tradition with some assurance of success. Simply following the guidelines alone will not guarantee a successful result. Nevertheless, these design rules can rarely be broken without reducing the impact of the bonsai specimen. Some key principles in bonsai aesthetics include:

- *Miniaturization*: By definition, a bonsai is a tree kept small enough to be container-grown while otherwise fostered to have a mature appearance.
- *Proportion among elements*: The most prized proportions mimic those of a full-grown tree as closely as possible. Small trees with large leaves or needles are out of proportion and are avoided, as is a thin trunk with thick branches.
- *Asymmetry*: Bonsai aesthetics discourage strict radial or bilateral symmetry in branch and root placement.
- *No trace of the artist*: The designer's touch must not be apparent to the viewer. If a branch is removed in shaping the tree, the scar will be concealed. Likewise, wiring should be removed or at least concealed when the bonsai is shown, and must leave no permanent marks on the branch or bark.
- *Poignancy*: Many of the formal rules of bonsai help the grower create a tree that expresses Wabi-sabi, or portrays an aspect of *mono no aware*.

Display

A bonsai display presents one or more bonsai specimens in a way that allows a viewer to see all the important features of the bonsai from the most advantageous position. That position emphasizes the bonsai's defined "front", which is designed into all bonsai. It places the bonsai at a height that allows the viewer to imagine the bonsai as a full-size tree seen from a distance, siting the bonsai neither so low that the viewer appears to be hovering in the sky above it nor so high that the viewer appears to be looking up at the tree from beneath the ground. Noted bonsai writer Peter Adams recommends that bonsai be shown as if "in an art gallery: at the right height; in isolation; against a plain background, devoid of all redundancies such as labels and vulgar little accessories."

For outdoor displays, there are few aesthetic rules. Many outdoor displays are semi-permanent, with the bonsai trees in place for weeks or months at a time. To avoid damaging the trees, therefore, an outdoor display must not impede the amount of sunlight needed for the trees on display, must support watering, and may also have to block excessive wind or precipitation. As a result of these practical constraints, outdoor displays are often rustic in style, with simple wood or stone components. A common design is the bench, sometimes with sections at different heights to suit different sizes of bonsai, along which bonsai are placed in a line. Where space allows, outdoor bonsai specimens are spaced far enough apart that the viewer can concentrate on one at a time. When the trees are too close to each other, aesthetic discord between adjacent trees of different sizes or styles can confuse the viewer, a problem addressed by exhibition displays.

Exhibition displays allow a large number of bonsai to be displayed in a temporary exhibition format, typically indoors, as would be seen in a bonsai design competition. To allow many trees to be located close together, exhibition displays often use a sequence of small alcoves, each containing one pot and its bonsai contents. The walls or dividers between the alcoves make it easier to view only one bonsai at a time. The back of the alcove is a neutral colour and pattern to avoid distracting the viewer's eye. The bonsai pot is almost always placed on a formal stand, of a size and design selected to complement the bonsai and its pot.

Indoors, a formal bonsai display is arranged to represent a landscape, and traditionally consists of the featured bonsai tree in an appropriate pot atop a wooden stand, along with a shitakusa (companion plant) representing the foreground, and a hanging scroll representing the background. These three elements are chosen to complement each other and evoke a particular season, and are composed asymmetrically to mimic nature. When displayed inside a traditional Japanese home, a formal bonsai display will often be placed within the home's tokonoma or formal display alcove. An indoor display is usually very temporary, lasting a day or two, as most bonsai are intolerant of indoor conditions and lose vigor rapidly within the house.

Containers

A variety of informal containers may house the bonsai during its development, and even trees that have been formally planted in a bonsai pot may be returned to growing boxes from time to time. A large growing box can house several bonsai and provide a great volume of soil per tree to encourage root growth. A training box will have a single specimen, and a smaller volume of soil that helps condition the bonsai to the eventual size and shape of the formal bonsai container. There are no aesthetic guidelines for these development containers, and they may be of any material, size, and shape that suit the grower.

Figure: *Assorted bonsai pots*

Completed trees are grown in formal bonsai containers. These containers are usually ceramic pots, which come in a variety of shapes and colours and may be glazed or unglazed. Unlike many common plant containers, bonsai pots have drainage holes in the bottom surface to complement fast-draining bonsai soil, allowing excess water to escape the pot. Growers cover the holes with a screening to prevent soil from falling out and to hinder pests from entering the pots from below. Pots usually have vertical sides, so that the tree's root mass can easily be removed for inspection, pruning, and replanting, although this is a practical consideration and other container shapes are acceptable.

There are alternatives to the conventional ceramic pot. Multi-tree bonsai may be created atop a fairly flat slab of rock, with the soil mounded above the rock surface and the trees planted within the raised soil. In recent times, bonsai creators have also begun to fabricate rock-like slabs from raw materials including concrete and glass-reinforced plastic. Such constructed surfaces can be made much lighter than solid rock, can include depressions or pockets for additional soil, and can be designed for drainage of water, all characteristics difficult to achieve with solid rock slabs. Other unconventional containers can also be used, but in formal bonsai display and competitions in Japan, the ceramic bonsai pot is the most common container.

For bonsai being shown formally in their completed state, pot shape, colour, and size are chosen to complement the tree as a picture frame is chosen to complement a painting. In general, containers with straight sides and sharp corners are used for formally shaped plants, while oval or round containers are used for plants with informal designs. Many aesthetic guidelines affect the selection of pot finish and colour.

For example, evergreen bonsai are often placed in unglazed pots, while deciduous trees usually appear in glazed pots. Pots are also distinguished by their size. The overall design of the bonsai tree, the thickness of its trunk, and its height are considered when determining the size of a suitable pot.

Some pots are highly collectible, like ancient Chinese or Japanese pots made in regions with experienced pot makers such as Tokoname, Japan or Yixing, China. Today many potters worldwide produce pots for bonsai.

Bonsai Styles

The Japanese tradition describes bonsai tree designs using a set of commonly understood, named styles. The most common styles include formal upright, informal upright, slanting, semi-cascade, cascade, raft, literati, and group/forest. Less common forms include windswept, weeping, split-trunk, and driftwood styles. These terms are not mutually exclusive, and a single bonsai specimen can exhibit more than one style characteristic. When a bonsai specimen falls into multiple style categories, the common practice is to describe it by the dominant or most striking characteristic.

A frequently used set of styles describes the orientation of the bonsai tree's main trunk. Different terms are used for a tree with its apex directly over the centre of the trunk's entry into the soil, slightly to the side of that centre, deeply inclined to one side, and inclined below the point at which the trunk of the bonsai enters the soil.

- Formal upright or *chokkan* style trees are characterized by a straight, upright, tapering trunk. Branches progress regularly from the thickest and broadest at the bottom to the finest and shortest at the top.
- Informal upright or *moyogi* trees incorporate visible curves in trunk and branches, but the apex of the informal upright is located directly above the trunk's entry into the soil line.
- Slant-style or *shakan* bonsai possess straight trunks like those of bonsai grown in the formal upright style. However, the slant style trunk emerges from the soil at an angle, and the apex of the bonsai will be located to the left or right of the root base.
- Cascade-style or *kengai* specimens are modelled after trees that grow over water or down the sides of mountains. The apex (tip of the tree) in the semi-cascade-style or *han kengai* bonsai extend just at or beneath the lip of the bonsai pot; the apex of a (full) cascade style falls below the base of the pot.

A number of styles describe the trunk shape and bark finish. For example, the deadwood bonsai styles identify trees with prominent dead branches or trunk scarring.

- Shari or *sharimiki* style involves portraying a tree in its struggle to live while a significant part of its trunk is bare of bark.

Although most bonsai trees are planted directly into the soil, there are styles describing trees planted on rock.

- Root-over-rock or *sekijoju* is a style in which the roots of the tree are wrapped around a rock, entering the soil at the base of the rock.
- Growing-in-a-rock or *ishizuke* style means the roots of the tree are growing in soil contained within the cracks and holes of the rock.

While the majority of bonsai specimens feature a single tree, there are well-established style categories for specimens with multiple trunks.

- Forest (or group) or *yose ue* style comprises a planting of several or many trees of one species, typically an odd number, in a bonsai pot.
- Multi-trunk or *ikadabuki* style has all the trunks growing out of one spot with one root system, and is actually a single tree.
- Raft-style or *netsuranari* bonsai mimic a natural phenomenon that occurs when a tree topples onto its side, for example, from erosion or another natural force. Branches along the top side of the trunk continue to grow as a group of new trunks.

Other Styles

A few styles do not fit into the preceding categories. These include:

- Literati or *bunjin-gi* style is characterized by a generally bare trunk line, with branches reduced to a minimum, and foliage placed towards the top of a long, often contorted trunk.
- Broom or *hokidachi* style is employed for trees with fine branching, like elms. The trunk is straight and branches out in all directions about 1/3 of the way up the entire height of the tree. The branches and leaves form a ball-shaped crown.
- Windswept or *fukinagashi* style describes a tree that appears to be affected by strong winds blowing continuously from one direction, as might shape a tree atop a mountain ridge or on an exposed shoreline.

Size Classifications

Japanese bonsai exhibitions and catalogues frequently refer to the size of individual bonsai specimens by assigning them to size classes. Not all sources agree on the exact sizes or names for these size ranges, but the concept of the ranges is well-established and useful to both the cultivation and the aesthetic understanding of the trees. A photograph of a bonsai may not give the viewer an accurate impression of the tree's real size, so printed documents may complement a photograph by naming the bonsai's size class. The size class implies the height and weight of the tree in its container.

In the very largest size ranges, a recognized Japanese practice is to name the trees "two-handed", "four-handed", and so on, based on the number of men required to move the tree and pot. These trees will have dozens of branches and can closely simulate a full-size tree. The very largest size, called "imperial", is named after the enormous potted trees of Japan's Imperial Palace. At the other end of the size spectrum, there are a number of specific techniques and styles associated solely with the smallest common sizes, *shohin* and *shito*. These techniques take advantage of the bonsai's minute dimensions and compensate for the limited number of branches and leaves that can appear on a tree this small.

Common names for bonsai size classesLarge bonsai

Common name	Size class	Tree Height
Imperial bonsai	Eight-handed	60–80 in (152–203 cm)
Hachi-uye	Six-handed	40–60 in (102–152 cm)
Dai	Four-handed	30–48 in (76–122 cm)
Omono	Four-handed	30–48 in (76–122 cm)
	Medium-size bonsai	
Common name	Size class	Tree Height
Chiu	Two-handed	16–36 in (41–91 cm)
Chumono	Two-handed	16–36 in (41–91 cm)
Katade-mochi	One-handed	10–18 in (25–46 cm)
	Miniature bonsai	
Common name	Size class	Tree Height
Komono	One-handed	6–10 in (15–25 cm)
Mame	One-handed	5–8 in (13–20 cm)
Shohin	Palm size	2–6 in (5–15 cm)
Shito	Fingertip size	2–4 in (5–10 cm)
Keshitsubo	Poppy-seed size	1–3 in (3–8 cm)

Indoor bonsai

Indoor bonsai are bonsai which are cultivated for the indoor environment. Traditionally, bonsai are temperate climate trees grown outdoors in containers. Kept in the artificial environment of a home, these trees weaken and die.

But a number of tropical and sub-tropical tree species will survive and grow indoors. Some of these are suited to bonsai aesthetics and can be shaped much as traditional outdoor bonsai are.

Note that bonsai and similar practices like *penjing*, *hòn non bΩ*, and *saikei* all involve the long-term cultivation of small trees in containers. The term *bonsai* is generally used in English as an umbrella term for all miniature trees in containers or pots. In this article *bonsai* should be understood to include any container-grown tree that is raised indoors and regularly styled or shaped, not just one being maintained in the Japanese bonsai tradition.

Indoor vs. Traditional Bonsai

The largest difference between indoor and traditional bonsai is, of course, the enjoyment of an attractive, fully leaved plant in winter instead of a dormant, leafless tree. Other differences include the faster growth rate of tropical plants, which accelerate all steps of the bonsai evolution. Moss covering, a common soil covering for outdoor bonsai, will not survive indoor conditions.

Bonsai vs. Other Forms of House Plant

Compared to the usual potted house plant, bonsai are rooted in a much smaller amount of soil. Consequently, they require more frequent watering and feeding.

This form is therefore best suited for drought-resistant species. Compared to usual house gardening, bonsais require a lot more pruning, both of branches and roots. This often requires a significant shift in attitude for house gardeners.

The spiritual benefits of bonsai cultivation (*bonsai no kokoro*) are available equally to classical and indoor bonsai gardeners.

Cultivation and Care

Many of the techniques used to grow and shape outdoor bonsai apply equally to indoor bonsai. For example, shaping techniques like pruning, wiring, and grafting work much the same for both types of bonsai. But indoor bonsai also have some special requirements for cultivation and care.

Lighting

An indoor room comfortably lit for human use provides too little light for most species of tree to grow. Few species will thrive with less than 500 lux available several hours per day. Successful indoor bonsai cultivation requires either selecting from the short list of low-light-tolerant trees, or providing additional lighting for the trees. Insufficient light may not kill some species outright, but will make their growth so slow that the bonsai shaping techniques cannot be used: the plant will not grow back after pruning or leaf trimming.

The simplest way to provide extra light is to place the bonsai close to an outside window. Care must be taken not to harm the plant either with cold air entering through the window, or with high radiated heat from direct exposure to the sun.

More controllable light, and more flexibility in situating the bonsai, can be obtained using artificial lighting. Fluorescent lighting (preferably with bulbs radiating growth-friendly spectra) and light-emitting diode lighting can give sufficient light to support a significant number of indoor bonsai species. These lights also have the advantage of blending with other indoor lighting, so that the plants can be grown in normal living quarters. For plants needing the highest amount of light, or for large numbers of bonsai being maintained indoors, a space dedicated to bonsai cultivation can support high-intensity lamps and optionally the special ventilation or cooling often needed for their use.

Temperature

Suitable temperatures for indoor bonsai extend from standard room temperature downward. The best temperature range for a tree is determined by its species and, less directly, by its normal range in the wild. Tropical trees are usually tolerant of room temperature year-round but cannot tolerate temperatures approaching freezing, as might occur near an open window in cold weather.

Semi-tropical and Mediterranean-climate trees often grow better when temperatures drop well below room temperature during winter months. The lower temperature, combined with shorter periods of daylight, triggers an expected annual dormancy which many non-tropical species need in order to thrive.

These non-tropical trees need a cool location in the winter, such as a cool window ledge or "cold room" area in the house. In addition to the need for seasonal temperature variation, many non-tropicals grow better when there is a distinct difference between day temperatures (warm) and night (cooler).

Humidity

Indoor conditions, particularly in homes outside the tropics, imply very low humidity. Both air conditioning and room heating reduce air humidity significantly. Some tree species, such as ficus sp. with their waxy leaves, are tolerant of low humidity. Many species, however, require additional humidification to survive indoors. As with lighting, some humidification solutions can be integrated into the regular living area and some require a dedicated space. Local humidity can be raised by locating the bonsai containers over a tray of water, or by placing them in a bed of dampened sand. For better effect, the bonsai can be placed in enclosures, such as terrariums, that are humidified from within. For the maximum humidity, a dedicated room with high-capacity evaporative or misting humidifiers can support the most sensitive bonsai.

Plants Suitable for Indoor Bonsai

Tropical

The creation of bonsai is limited only by the imagination and talent of the gardener, although some species are much more suitable than others. Members of the genus *Ficus* are among the most versatile, while many succulents can be grown in a similar fashion. Here is an incomplete list of the most popular species.

- *Crassula ovata*: The jade plant is a very robust and drought-resistant house plant. The miniature cultivars like the baby jade plant (*C. ovata arborescens*) is considered the best plant for a first bonsai. This plant will sprout on old wood. Thus, an old specimen can be pruned back to a stockier shape with thick trunk. It is kept dry in winter, placed outdoors in summer for full growth. Its roots are thin and cannot be exposed.
- *Cycas revoluta*: This plant is often used as an indoor bonsai. Though it does not have a traditional tree shape, it can give an image of a tropical island.
- *Dracaena marginata*: The dragon plant has an interesting palm-like shape. It can sprout on old wood. It does not tolerate root exposure.
- *Ficus benjamina*: The Weeping Fig is a popular indoor tree that lends itself to the classical, upright form. It is one of the few tropicals that are accepted as "true" bonsai. The miniature cultivars like 'Too Little' are also well suited for bonsai. It forms aerial roots and can be shaped as a banyan tree. Ficus are

intolerant to branch down-pruning; one must start with a small tree and keep it small. They are sensitive to stress.

- *Ficus neriifolia*: According to Jerry Meislik, "the most useful fig for bonsai is the willow leafed fig. The small leaf is in excellent scale for bonsai and the tree has good branch ramification, good basal rootage and excellent aerial root formation."
- *Portulacaria afra*: The dwarf jade looks a lot like a baby jade plant and is used similarly.
- *Schefflera arboricola*: The Hawaiian umbrella tree is a popular, hardy houseplant that is ideal for irregular, banyan or roots-on-rock forms. Since it can sprout on old wood, an old specimen can be pruned back to a stockier shape with thick trunk and roots. It tolerates root exposure very well, is drought-resistant and requires a moderate amount of light. Under high humidity conditions, it produces aerial roots and can therefore be shaped as a banyan tree.
- *Schlumbergera*: the Christmas cactus does not have a real trunk but easily lends itself to a cascade-type bonsai shape. It tolerates shade, not drought.
- *Tamarindus indica* The tamarind has recently become popular in bonsai culture, frequently used in Asian countries like Indonesia, Taiwan and the Philippines. In the last Japan Airlines World Bonsai competition, Mr. Budi Sulistyo of Indonesia won the second prize with an ancient tamarind bonsai.

Small succulents may be used as accent plants:

- Rhipsalis (Hatiora) salicornioides

Other Climatic Origins

With proper care, a number of non-tropical plants can also thrive as indoor bonsai.

- *Araucaria heterophylla*: The Norfolk Pine is a needle-bearing evergreen. It does well indoors and is often seen in commercial buildings, but is not easily shaped to bonsai styles.
- *Ardisia crenata*: The spear-flower will flower and produce berry-like fruit when grown indoors.
- *Bougainvillea*: This flowering shrub, native to South America, will grow indoors with sufficient light.
- *Bucida molineti*: The black olive is native to southern Florida in the USA, and has a compact growth habit suited to bonsai forms.

- *Boxwood*: Found on nearly every continent, boxwood varieties are tough but attractive bonsai. The varieties *Buxus harlandii* and *Buxus microphylla* (Kingsville boxwood) have tiny leaves and develop mature-looking bark, supporting bonsai aesthetics.
- *Calliandra haematocephala* (Red powderpuff): The *Calliandra* family includes evergreen bushes growing in tropical and subtropical areas worldwide. The *haematocephala* species has small needle-like leaves and unusual-looking feathery blossom that appear in spring and summer.
- *Camellia japonica* (Common camellia), *Camellia sinensis*: In the mountains of Japan and Korea, trees of the *Camellia* family grow up to 10 to 12 metres high. *Camellia sinensis* in particular suit bonsai treatment, with relatively small leaves and flowers.
- *Carmona microphylla*: The sturdy trunk and glossy green leaves of the Fukien Tea make an attractive indoor bonsai. Tiny white flowers appear regularly, and a characteristically rough or gnarled bark gives an appearance of age on even young bonsai.
- *Casuarina equisetifolia* (Horsetail tree): The horsetail tree grows up to 25 metres in the wild, but can be cultivated indoors as a compact bonsai. Its foliage is needle-like although the plant is considered a deciduous tree, and it can be shaped to resemble a spruce or pine.
- *Citrus japonica 'Hindsii'* (Kumquat): This fruiting tree is one of the kumquat varieties, and relatives with small leaves suitable for indoor bonsai include *Citrus japonica 'Margarita'*, *Citrus japonica 'Japonica'* (Marumi kumquat), × *Citrofortunella mitis*, *Citrus aurantifolia* (Key lime), and *Citrofortunella microcarpa* (Calamondin). The trees flower and may even bear fruit in pot cultivation.
- *Cupressus macrocarpa*: The Monterey Cypress is one of the few evergreen conifers that can be grown indoors. Trunk thickness is difficult to develop, but the fine juniper-like foliage suits even small bonsai.
- *Feroniella lucida*: Native to southeast Asia, this tree flowers and produces edible fruit. It is a slow grower in a container, but small leaves and a rough trunk make it very suitable for bonsai.
- *Fraxinus uhdei* (Tropical ash): Though the leaf clusters do not match common bonsai aesthetics, this tree can be shaped into an effective bonsai, particularly as the trunk ages and develops character.

- *Ligustrum sp.*: Some privet species, including *Ligustrum japonica* and *Ligustrum rotundifolium*, are suitable subjects for indoor bonsai.
- *Serissa*: This is a delicate flowering plant with tiny leaves that can grow indoors year round.
- *Syzygium paniculatum*, syn. *Eugenia paniculata*: A flowering and fruiting evergreen tree with slender leaves, this native of Australia grows to 6 metres in the wild. As a bonsai, it can be shaped in most styles suiting deciduous trees.
- *Ulmus parvifolia*: The Chinese elm is often used as outdoor bonsai, and it can be raised indoors as well.

Trees Physiological and Biochemical Processes

Forest Pathology

Forest pathology is the research of both biotic and abiotic maladies affecting the health of the a forest or tree, primarily fungal pathogens and their insect vectors. It is a subfield of forestry and plant pathology. Pathology is from Ancient Greek ðÜèïò, pathos which may be translated into English as either "experience" or "suffering". and -ëïãßα, -logia, "An account of" or "the study of".

Forest pathology is part of the broader approach of forest protection.

Abiotic Factors

There a number of abiotic factors which affect the health of a forest such as moisture issues like drought, winter-drying, waterlogging or Precipitation such as hail, snow, storm.

Wind is also an important abiotic factor as Windthrow (the uprooting or breaking of trees due to high winds) causes an obvious and direct loss of stability to a forest or its trees.

Often, abiotic factors and biotic factors will affect a forest at the same time. For example, if wind speed is 80 km per hour then many trees which have root rot (caused by a pathogen) are likely to be thrown. Higher wind speeds are necessary to damage healthier trees (such as 125 km per hour).

Fire and related abiotic factors also affect the health of forest. Lightning. Nutrient imbalances: deficiencies, chemicals (toxic salts, herbicides, air pollutants). Temperature: low, high.

The effects of man often alter a forest's predisposition to damage from both abiotic and biotic effects. For example soil properties may be altered by heavy machinery.

Biotic Factors

- Fungi: Ascomycota, Basidiomycota and fungi imperfecti
- Oomycota: Phytophthora.
- Bacteria.
- Phytoplasmas.
- Viruses.

"Insects"

There is a category listing insect pests of temperate forests in Wikipedia

- Ips (genus) bark beetles
- Bark beetle
- Ambrosia beetle
- Cerambycidae
- Black Arches

Some of these factors act in concert (well all do to a degree). For example Amylostereum areolatum which is spread by theSirex woodwasp. The fungus gains access to new trees to live off and the woodwasp larvae gain food.

Parasitic Flowering Plants

- Many plants can parasitize trees via root to root contact. Many of these parasitic plants originate in the tropical and subtropical climates.

Animals

Nematodes, insects especially bark beetles, mammals may browse. Browsing can be prevented with Tree shelters

Part of forest pathology is forest entomology. Forest entomology encompasses the study of all insects and arthropods such as mites, centipedes and millipedes which live in a and interact in forest ecosystems and the management of insect pests that cause the degrading, defoliation, crown die-back or death of trees.

Thus the scope is wide and includes:

- Documentation of all insect species and related arthropoda in natural and man-made forests, and the study and ecology of those species.
- Description and assessment of damage to tree structures (parts of a tree), to forest stands, landscape effects and to wood products, timber in service and other ecosystem services.

- Eradication of recently introduced pests, or long-term management of established exotics and indigenous pests, to minimise losses in wood quality and wood production, and to reduce tree mortality.
- Assessments of forest operations, or of management impacts, on the invertebrate fauna, and the alleviation of any adverse effects on these invertebrates.

Hazard Trees

- The likelihood of property damage or personal injury due to tree failure. Hazard includes not only the tree's condition, but the potential target as well. Rating systems, procedures and guidelines have been developed for decision making but knowledge, judgement, and experience are an important part to the process.

Pathogens that Affect Trees

There is a category listing tree diseases in Wikipedia.

- Armillaria sp of which cause white rot root disease
- Ash dieback
- Cenangium
- Hymenoscyphus pseudoalbidus which causes as Ash Dieback
- Heterobasidion annosum which causes Annosum or red root rot, the economically most significant pathogen in the Northernn hemisphere.
- Chestnut blight
- Rickettsia which causes possibly this citrus greening disease
- Spiroplasma
- Dutch elm disease
- Ink disease
- Emerald ash borer
- Olive tree pathology
- Witch's broom
- White Pine Blister Rust
- Phytophthora cinnamomi a root rot
- Phytophthora ramorum which causes "Sudden oak death"
- Polypore or Bracket fungus
- Tinder conk

Signs and Symptoms

Symptoms are a result of a pathogen:

- Blight
- Burl
- Canker
- Chlorosis
- Drunken trees
- Forest dieback
- Gall
- Girdling
- Leaf scorch
- Root rot
- Virescence
- Wilt disease

Signs are the visible presence of a part of a pathogen:

- Ascus is a part of an acomycota fungus.
- Conk (fungi) is the fruiting body of a bracket fungus.
- Hypha are collectively called a mycelium
- Mycelial cord or rhizomorphs

Pathology Detection

This can be done by dogs or machines smelling the trees. It is similar to noses used to find truffles. It can also be done by monitoring and identification can happen via tree clinics, experts such as arborists or even non-experts through citizen science. It is important to consider the disease triangle when evaluating pathologies. Demonstration of suspected active agents can be done by confirmation of Koch's postulates.

Forest Integrated Pest Management

Forest integrated pest management or Forest IPM is the practice of monitoring and managing pest and environmental information with pest control methods to prevent pest damage to forests and forest habitats by the most economical means. Forest IPM practices vary from region to region and particularly by state, according to the habitat and forests present.

In the state of Vermont, two common pests are of particular significance, the Hemlock Woolly Adelgid or HWA, and thrips.

8

Multistoried Agroforestry Systems Studies

Because of limited land and high population densities, farmers have developed highly sophisticated multistoried tree cropping systems. The majority of rural people in Bangladesh rely on their homesteads, rather than on natural forests for a range of products and services. This situation presents an interesting contrast for forestry professionals as they are seen as providing technical advice to maximize yield and better market income generating tree species, rather than enforcing strict policies on extraction and management of forest products.

Multi-storied cropping systems and agro-forestry systems help to reduce the impacts of hazards such as floods and droughts that lead to land degradation, while meeting the needs of local people for food, firewood, fodder and timber.

Under the multi-storied cropping, crops underneath the tall trees or horticultural crops withstand shading effects and provide micro-climatic conditions needed by all the crops. This system has at least three layers of intercropped plants of different heights. It is commonly developed by small farmers, as they:

- Plant nurse trees for shade-tolerant crops like coffee and cacao
- Make productive use of space between and under cash crops
- Enrich fallows with species like yams
- Establish a variety of low input, low maintenance tree crops.

Objective: The objective of the technology is to grow the best combination and interface between crops and trees for control of land degradation, increasing system productivity and meeting domestic needs of food, fuel, fodder and timber of rural people.

Implementation of the Technology: Before initiating, it is essential to find the best combination possible, for this, an ecological and economic planning should be undertaken, considering the following:

a. Ecological planning: For any crop combination, predict how the multi storied agro-forestry system will grow and change over time. Each year, some plants will become less productive, and others will grow and cast additional shade. Plan the development of the multi storied agro forestry system so that:

    i. The first plants established are adapted to full sunlight

    ii. Plants that can give shade are planted before plants that can tolerate or require shade.

    iii. Plants that enrich the soil are planted before plants that give shade.

    iv. Plants that require sunlight are not planted where other plants will shade them before they mature

    v. Medium-sized and large trees will have room to grow and will not become too crowded later. Think ahead to the width of the tree's canopy when it is mature. Consider whether the trees around it will also grow tall and wide, if they will crowd it, they should be harvested or not planted at all.

    vi. All growing space is used: crops fit together vertically (tall, medium and short), horizontally (al planting spots occupied), and underground (deep-rooted and shallow-rooted plants.)

b. Economic planning: Any farming system requires planning to diversify products and spread labour and income through the year. The first crops planted should provide food or income within 3-4 months (for example, sweet potato). Choose crops that give both subsistence produce and cash income. Plant trees that bear different fruits at different times of the year. A multi-storied agro-forest will change every year for the first several years, with changes in labour requirements and products.

Steps to implement a multi-stories agro forestry system: Plant nurse trees: Fast-growing nitrogen-fixing trees can be planted first to improve site conditions. This approach is often used for crops like coffee on poor sites, where nurse trees are valued for shade and soil enrichment throughout the life of the coffee crop. Common species suitable for this method are Albiia lebbek (chres), Gliricidia sepium (snaov), Dalbergia cochinchinensis (kra ngoung), Pterocarpus indicus (thnong) and Acacia mangium. All these species have special characteristics such as: they cast a light shade or can be pruned to adjust shade; are nitrogen-fixing; produce timber, fodder or other useful products; are sun-loving; adapted to local climates and soils; and fast growing.

Intercrop annual crops: Choose an intercrop each planting season that is adapted to current sun or shade conditions. Shade-tolerant under-storey crops include: Ananas cosmosus (pineapple), Capsium spp. (peppers), Colocasia esculenta (sweet taro), Curcuma domestica (turmeric), Anthurium spp. (flowers), Ipomea batatas (sweet potato), Xanthosoma sagittifolium (dryland taro), Zingiber officinarum (ginger) and mushroom cultures.

Plant shade-tolerant trees: After one year or when the nurse trees provide enough shade, plant shade-tolerant trees. Plant them 3-4 m apart, or at the recommended spacing for the tree crop. Plant them 1-2 m away from the nurse trees.

Fertilize, prune and weed as for other tree crops.

Prune nurse trees: When nurse trees provide too much hade, cut branches and use the leaves as mulch.

Thin trees: Watch for adjacent trees with canopies starting to grow into each other. Also watch out for sun-loving trees to make space for healthy or higher –value trees. Be careful not to damage other trees when cutting. The remaining trees will grow faster.

Enrich with other crops. Plant shade-tolerant crops for food and cash income.

Example of crop combinations in the mid-hills region in Nepal

Under multistoried cropping, coffee and pineapple are the major crops being grown as shade loving crops under tall trees in mid-hills (eg. Arghakhanchi and Udaipur districts). Other crops taken under this system are normally citrus, maize, legumes, vegetables and finger millet for lower altitude. At higher altitudes, crops taken under this system are apple, walnut, maize, beans and finger millet. This multi-storied cropping system will reduce the impacts of hazards like high intensity rainfall, soil erosion and landslides.

Under agro-forestry or agro-silvi-pastoral system, single, double or multiple crop and/or grass species such as single or double crops of maize, millet or grasses are planted between rows of tree species (e.g. *Delbergia sisso*) under various tree-crop intercropping systems. Some good examples of tree-crop intercropping practised in the hills are Uttis (Alnus nepalensis) /cardamom intercropping for more productivity and benefits. Community agro-forestry system implemented and regulated by users' group are the examples for a successful model. Agricultural lands below settlements/forest areas offer good agro-ecosystem by providing leaf litters and organic matter for agricultural production below. Thus both the multi-storied and agro-forestry systems lead to efficient utilization of resources, soil conservation, and minimizing risks of crop yield loss thereby increasing household income and employment opportunity.

Silviculture

Silviculture is the practice of controlling the establishment, growth, composition, health, and quality of forests to meet diverse needs and values. The name comes from the Latin *silvi-* (forest) + culture (as in growing). The study of forests and woods is termed silvology. Silviculture also focuses on making sure that the treatment(s) of forest stands are used to preserve and to better their productivity.

To some the distinction between forestry and silviculture is that silviculture is applied at the stand level and forestry is broader. For example John D. Matthews says "complete regimes for regenerating, tending, and harvesting forests" are called "silvicultural systems".

So active management is required for silviculture, whereas forestry can be natural, conserved land without a stand level treatment being applied. A common taxonomy divides silviculture into regenerating, tending and harvesting techniques.

Regeneration

Forest regeneration is the act of renewing tree cover by establishing young trees, generally promptly after the previous stand or forest has been removed. The method, species, and density are chosen to meet the goal of the landowner.

It may be divided into natural regeneration: "Human-assisted natural regeneration" means establishment of a forest age class from natural seeding or sprouting in an area after harvesting in that area through selection cutting, shelter (or seed-tree) harvest, soil preparation, or restricting the size of a clear-cut stand to secure natural regeneration from the surrounding trees.

The process of natural regeneration involves the renewal of forests by means of self-sown seeds, root suckers, or coppicing. In natural forests, conifers rely almost entirely on regeneration through seed.

Most of the broadleaves, however, are able to regenerate by the means of emergence of shoots from stumps (coppice) and broken stems.

Tree provenance is important in artificial regeneration. Good provenance takes into account suitable tree genetics and a good environmental fit for planted / seeded trees in a forest stand. The wrong genotype can lead to failed regeneration, or poor trees that are prone to pathogens and undesired outcomes.

Artificial regeneration has been a more common method involving planting because it is more dependable than natural regeneration. Planting can involve using seedlings (from a nursery), (un)rooted cuttings, or seeds.

Whichever method is chosen it can be assisted by tending techniques also known as intermediate stand treatments.

Tending

Enrichment Planting

A strategy for enhancing natural forests' economic value is to increase their concentration of economically important, indigenous tree species by planting seeds or seedlings for future harvest, which can be accomplished with enrichment planting (EP). This means increasing the planting density (i.e., the numbers of plants per hectare) in an already growing forest stand."

Release Treatments

- ***Weeding:*** A process of getting rid of samplings' or seedlings' competition by mowing, application of herbicide, or other method of removal from the surroundings.
- ***Cleaning:*** Release of select saplings from competition by overtopping trees of a comparable age. The treatment favours trees of a desired species and stem quality.
- ***Liberation Cutting:*** A treatment that releases tree seedling or saplings by removing older overtopping trees.

Thinning

Thinning is an operation that artiûcially reduces the number of trees growing in a stand with the aim of hastening the development of the remainder. The goal of thinning is to control the amount and distribution of available growing space. By altering stand density, foresters can influence the growth, quality, and health of residual trees. It also provides an opportunity to capture mortality and cull the commercially less desirable, usually smaller and malformed, trees.

Unlike regeneration treatments, thinnings are not intended to establish a new tree crop or create permanent canopy openings.

Common Thinning Methods

- ***Low Thinning:*** method of what they do is what they predicted to happen naturally. It emphasizes crown class.
- Crown Thinning (*thinning from above* or *French method*)
- Selection Thinning (*thinning of dominants* or *Borggreve method*). It focuses primarily on quality.
- Mechanical Thinning (*row thinning* or *geometric thinning*)
- Free Thinning
- Loose Thinning

Previous studies have demonstrated that repeated thinnings over the course of a forest rotation increase carbon stores relative to stands that are clear-cut on short rotations and that the carbon benefits differ according to thinning method (e.g., thinning from above versus below).

Ecological thinning is where the primary aim of forest thinning is to increase growth of selected trees, favouring development of wildlife habitat (such as hollows) rather than focusing on increased timber yields. Ecological thinning can be considered a new approach to landscape restoration for some types of eucalypt forests and woodlands in Australia.

Pruning

Pruning, as a silvicultural practice, refers to the removal of the lower branches of the young trees (also giving the shape to the tree) so clear knot-free wood can subsequently grow over the branch stubs. Clear knot-free lumber has a higher value. Pruning has been extensively carried out in the Radiata pine plantations of New Zealand and Chile, however the development of Finger joint technology in the production of lumber and mouldings has led to many forestry companies reconsidering their pruning practices. "Brashing" is an alternative name for the same process. Pruning can be done to all trees, or more cost effectively to a limited number of trees. There are two types of pruning: natural or self-pruning and artificial pruning. Most cases of self-pruning happen when branches do not receive enough sunlight and die. Wind can also take part in natural pruning which can break branches. Artificial pruning is where people are paid to come and cut the branches. Or it can be natural, where trees are planted close enough that the effect is to cause self-pruning of low branches as energy is put into growing up for light reasons and not branchiness.

Forest Care

This may be reducing browsing. It can also be techniques for dealing with forest pathogens.

Common Methods

Silvicultural regeneration methods combine both the harvest of the timber on the stand and re-establishment of the forest. The proper practice of sustainable forestry should mitigate the potential negative impacts, but all harvest methods will have some impacts on the land and residual stand. The practice of sustainable forestry limits the impacts such that the values of the forest are maintained in perpetuity. Silvicultural prescriptions are specific solutions to a specific set of circumstances and management objectives. Following are some common methods:

- Single-tree selection - The single-tree selection method is an uneven-aged regeneration method most suitable when shade tolerant species regeneration is desired. It is typical for older and diseased trees to be removed, thus thinning the stand and allowing for younger, healthy trees to grow. Single-tree selection can be very difficult to implement in dense or sensitive stands and residual stand damage can occur. This method is also disturbs the canopy layer the least out of all other methods.
- Group selection - The group selection method is an uneven-aged regeneration method that can be used when mid-tolerant species regeneration is desired. The group selection method can still result in residual stand damage in dense stands, however directional falling can minimize the damage. Additionally, foresters can select across the range of diametre classes in the stand and maintain a mosaic of age and diametre classes.
- Clearcutting - An even-aged regeneration method that can employ either natural or artificial regeneration. It involves the complete removal of the forest stand at one time. Clearcutting can be biologically appropriate with species that typically regenerate from stand replacing fires or other major disturbances, such as Lodgepole Pine (*Pinus contorta*). Alternatively, clearcutting can change the dominating species on a stand with the introduction of non-native and invasive species as was shown at the Blodgett Experimental Forest near Georgetown California. Additionally, clearcutting can prolong slash decomposition, expose soil to erosion, impact visual appeal of a landscape and remove essential wildlife habitat. It is particularly useful in regeneration of tree

species such as Douglas-fir (*Pseudotsuga menziesii*) which is shade intolerant.. In addition, the general public's distaste for even-aged silviculture, particularly clearcutting, is likely to result in a greater role for uneven-aged management on public lands as well. Across Europe, and in parts of North America, even-aged, production-orientated and intensively managed plantations are beginning to be regarded in the same way as old industrial complexes: something to abolish or convert to something else.

- Patch cut
- Seed-tree - An even-aged regeneration method that retains widely spaced residual trees in order to provide uniform seed dispersal across a harvested area. In the seed-tree method, 2-12 seed trees per acre (5-30/ha) are left standing in order to regenerate the forest. They will be retained until regeneration has become established at which point they may be removed. It may not always be economically viable or biologically desirable to re-enter the stand to remove the remaining seed trees. Seed-tree cuts can also be viewed as a clearcut with natural regeneration and can also have all of the problems associated with clearcutting. This method is most suited for light-seeded species and those not prone to windthrow.
- Shelterwood - In general terms, the shelterwood system is a series of partial cuts that removes the trees of an existing stand over several years and eventually culminates in a final cut that creates a new even-aged stand. Its an even-aged regeneration method that removes trees in a series of three harvests: 1) Preparatory cut; 2) Establishment cut; and 3) Removal cut. The success of practising a shelterwood system is closely related to: 1. the length of the regeneration period, i.e. the time from the shelterwood cutting to the date when a new generation of trees has been established; 2.the quality of the new tree stand with respect to stand density and growth; and 3.the value increment of the shelter trees. Information on the establishment, survival and growth of seedlings influenced by the cover of shelter trees, as well as on the growth of these trees, is needed as a basis for modelling the economic return of practising a shelterwood system. The method's objective is to establish new forest reproduction under the shelter of the retained trees. Unlike the seed-tree method, residual trees alter understory environmental conditions (i.e. sunlight, temperature, and moisture) that influence tree seedling growth. This method can also find a middle ground with

the light ambiance by having less light accessible to competitors while still being able to provide enough light for tree regeneration. Hence, shelterwood methods are most often chosen for site types characterized by extreme conditions, in order to create a new tree generation within a reasonable time period. These conditions are valid foremost on level ground sites which are either dry and poor or moist and fertile.

- Coppicing - A regeneration method which depends on the sprouting of cut trees. Most hardwoods, the coast redwood, and certain pines naturally sprout from stumps and can be managed through coppicing. Coppicing is generally used to produce fuelwood, pulpwood, and other products dependent on small trees. A close relative of coppicing is pollarding. Three systems of coppice woodland management are generally recognized: simple coppice, coppice with standards, and the coppice selection system.
  - In Compound coppicing or coppicing with standards, some of the highest quality trees are retained for multiple rotations in order to obtain larger trees for different purposes.
- Variable retention - A harvesting and regeneration method which is a relatively new silvicultural system that retains forest structural elements (stumps, logs, snags, trees, understory spieces and undisturbed layers of forest floor) for at least one rotation in order to preserve environmental values associated with structurally complex forests.

"Uneven-aged and even-aged methods differ in the scale and intensity of disturbance. Uneven-aged methods maintain a mix of tree sizes or ages within a habitat patch by periodically harvesting individual or small groups of trees, Even-aged methods harvest most or all of the overstory and create a fairly uniform habitat patch dominated by trees of the same age". Even-aged management systems have been the prime methods to use when studying the effects on birds.

Ecological Thinning

Ecological thinning is a silvicultural technique used in forest management that involves cutting trees to improve functions of a forest other than timber production.

Although thinning originated as a man-made forest management tool, aimed at increasing timber yields, the shift from production forests to multifunctional forests brought with it the cutting of trees to manipulate an ecosystem for various reasons, ranging from removing non-native species from a plot to removing poplars growing on a riverside beach aimed at recreational use. Since the 1970s, leaving the thinned trees on the forest floor has become an increasingly common policy: wood can be decomposed in a more natural fashion, playing an important role in increasing biodiversity by providing habitat to various invertebrates, birds and small mammals. Many fungi (e.g. *Calocera viscosa*) and mosses are saproxylic or epixylic as well (e.g. *Marchantiophyta*) – some moss species completing their entire life-cycle on a single log.

Where trees are managed under a commercial regime, competition is reduced by removing adjacent stems that exhibit less favourable timber quality potential. When left in a natural state trees will "self-thin", but this process can be unreliable in some circumstances. Examples of this can be found in the Buxus - Ironbark forests and woodlands of Victoria (Australia) where a large proportion of trees are coppice, resultant from timber cutting in decades gone by.

Ecological Thinning Research

Research programmes under way in various parts of the world (e.g. USA and Australia) are aimed at providing an alternate approach in forest management where conservation objectives are a high priority. Methods of ecological thinning being developed on silvicultural techniques for local forest types. Ecological thinning is being developed using two principles: 1. appropriate stem reduction to reduce competition and 2. retention of trees (selection) that are more suitable for wildlife (i.e. not timber production). An example of ecological thinning research is the project in Victoria's Box-Ironbark forests, investigating various thinning and timber removal methods under an Adaptive Management or AEM framework. The primary objective is to generate (over time) a number of forest habitat values (i.e. tree hollows) that are crucial for wildlife conservation.

Enzymes Activity and IAA Contents in Soils

Soil microorganisms can be closely associated with soil animals such as earthworms for the delivery of ecosystem functions such as organic matter mineralization. The release of nutrients due to this dual interaction has consequences on plant growth. In addition to nutrients, interdependence between microorganisms, plants, and earthworms has led to the selection of diversified signal molecules, from low-molecular weight compounds such as auxin (e.g., IAA), amino acids, and secondary metabolites, to high-molecular weight compounds such as proteins and polysaccharides. Emission of signal molecules by microorganisms can be deleterious for plants in the case of pathogenic bacteria and invertebrate herbivores, or positive with mycorrhizas, nitrogen-fixing symbiotic bacteria, or free living bacteria such as plant growth promoting rhizobacteria (PGPR). Auxin-like compounds have also been detected in earthworm dejections, and could be responsible for the positive effect of earthworms and vermicompost on plant growth. The same molecule can have positive or negative effects according to its concentration and plant sensitivity.

Indole-3-acetic Acid

Indole-3-acetic acid (IAA) is the most common, naturally-occurring, plant hormone of the auxin class. It is the best known of the auxins, and has been the subject of extensive studies by plant physiologists. Chemically, IAA is a carboxylic acid in which the carboxyl group is attached through a methylene group to the C-3 position of an indole ring. In appearance, IAA is a colourless solid.

Biosynthesis and Biological Activity

IAA is predominantly produced in cells of the apex (bud) and very young leaves of a plant. Plants can synthesize IAA by several independent biosynthetic pathways. Four of them start from tryptophan, but there is also a biosynthetic pathway independent of tryptophan. Plants mainly produce IAA from tryptophan through indole-3-pyruvic acid. IAA is also produced from tryptophan through indole-3-acetaldoxime in *Arabidopsis*.

IAA has many different effects, as all auxins do, such as inducing cell elongation and cell division with all subsequent results for plant growth and development. On a larger scale, IAA serves as signaling molecule necessary for development of plant organs and coordination of growth. It is formed from tryptophan, along with tryptophol in rats infected by *Trypanosoma brucei gambiense*.

Synthetic Analogs

There are less expensive and metabolically stable synthetic auxin analogs on the market for use in horticulture, such as indole-3-butyric acid (IBA) and 1-naphthaleneacetic acid (NAA).

Studies of IAA in the 1940s led to the development of the phenoxy herbicides 2,4-dichlorophenoxyacetic acid (2,4-D) and 2,4,5-trichlorophenoxyacetic acid (2,4,5-T). Like IBA and NAA, 2,4-D and 2,4,5-T are metabolically and environmentally more stable analogs of IAA. However, when sprayed on broad-leaf dicot plants, they induce rapid, uncontrolled growth, eventually killing them. First introduced in 1946, these herbicides were in widespread use in agriculture by the middle of the 1950s.

Synthesis

Chemically, it can be synthesized by the reaction of indole with glycolic acid in the presence of base at 250 °C:

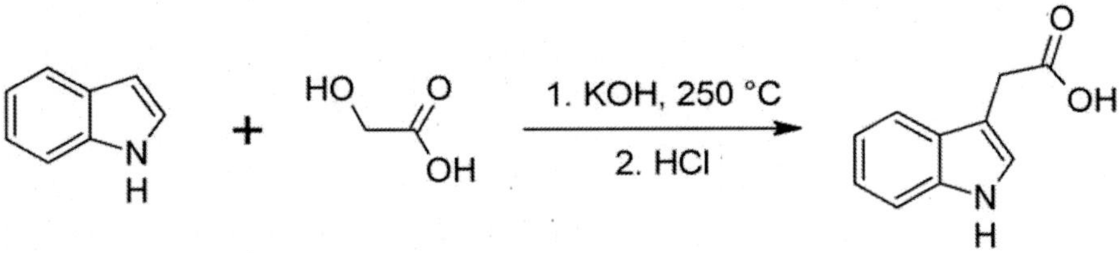

Many methods for its synthesis have been developed since its original synthesis from indole-3-acetonitrile.

Toxicity on Human Organs

Little research has been conducted on the effects of IAA on humans and toxicity data is limited. No data on carcinogenic, teratogenic, or developmental effects have been created. IAA is listed in its MSDS as potentially mutagenic to mammalian somatic cells. It is listed as a potential skin, eye, and respiratory irritant, and users are warned not to ingest it. Protocols for ingestion, inhalation, and skin/eye exposure are standard for moderately poisonous compounds and include thorough rinsing in the case of skin and eyes, fresh air in the case of inhalation, and immediately contacting a physician in all cases to determine the best course of action. Do not induce vomiting in the case of ingestion nor attempt to give anything by mouth to an unconscious person. The NFPA 704 health hazard rating for IAA is 2, which denotes a risk of temporary incapacitation with intense or prolonged, but not chronic exposure, and a possibility of residual injury.

Other Uses

Effects on Embryonic Rat Brain Development

Studies have been performed to determine the effects IAA and similar chemicals on the development of the cerebral cortex during early embryogenesis. One study performed in rats determined that such compounds decreased the locomotor activities of the embryos/ fetuses. Further, treatment with IAA and analog 1(methyl)-IAA resulted in significantly decreased brain sizes (relative to body weigh)in embryonic day 21 animals.

Cancer

IAA and horseradish peroxidase (HRP) have been proposed on the basis of polymer transfer and gene-direction, they may be useful in cancer therapy. Radical-IAA IAA in the human body will attach to cells with that characterized by HRP. By this means the HRP reaction cells can be selectively killed. But because of the competitive bonding of the IAA-HRP more studies must be done before clinical studies can begin.

Auxin

Auxins are a class of plant hormones (or plant growth substances) with some morphogen-like characteristics. Auxins have a cardinal role in coordination of many growth and behavioural processes in the plant's life cycle and are essential for plant body development. Auxins and their role in plant growth were first described by the Dutch scientist Frits Warmolt Went. Kenneth V. Thimann isolated this phytohormone and determined its chemical structure as indole-3-acetic acid.

Native Auxins

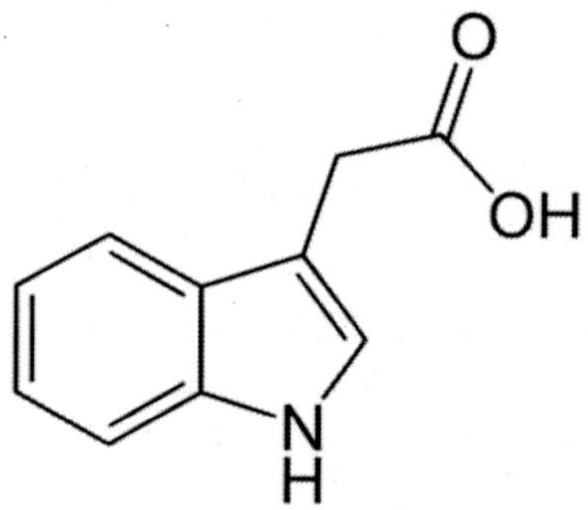

Indole-3-acetic acid (IAA) is the most abundant and the basic auxin natively occurring and functioning in plants. It generates the majority of auxin effects in intact plants, and is the most potent native auxin.

There are three more native — endogenous auxins All auxins are compounds with aromatic ring and a carboxylic acid group:

Auxins derive their name from the Greek word *auxein* - "to grow/ increase". They were the first of the major plant hormones to be discovered.

The (dynamic and environment responsive) pattern of auxin distribution within the plant is a key factor for elongation during growth of the stem and root, its reaction to its environment, and specifically for development of plant organs (such as leaves or flowers). It is achieved through very complex and well coordinated active transport of auxin molecules from cell to cell throughout the plant body — by the so-called polar auxin transport. Thus, a plant can (as a whole) react to external conditions and adjust to them, without requiring a nervous system. Auxins typically act in concert with, or in opposition to, other plant hormones. For example, the ratio of auxin to cytokinin in certain plant tissues determines initiation of root versus shoot buds.

On the molecular level, all auxins are compounds with an aromatic ring and a carboxylic acid group. The most important member of the auxin family is indole-3-acetic acid (IAA). IAA generates the majority of auxin effects in intact plants, and is the most potent native auxin. And as native auxin, its stability is controlled in many ways in plants, from synthesis, through possible conjugation to degradation of its molecules, always according to the requirements of the situation. However, molecules of IAA are chemically labile in aqueous solution, so it is not used commercially as a plant growth regulator.

- The four naturally occurring (endogenous) auxins are IAA, 4-chloroindole-3-acetic acid, phenylacetic acid and indole-3-butyric acid; only these four were found to be synthesized by plants. However, most of the knowledge described so far in auxin biology and as described in the article below, apply basically to IAA; the other three endogenous auxins seems to have rather marginal importance for intact plants in natural environments. Alongside endogenous auxins, scientists and manufacturers have developed many synthetic compounds with auxinic activity.
- Synthetic auxin analogs include 1-naphthaleneacetic acid, 2,4-dichlorophenoxyacetic acid (2,4-D), and many others.

Some synthetic auxins, such as 2,4-D and 2,4,5-trichlorophenoxyacetic acid (2,4,5-T), are used also as herbicides. Broad-leaf plants (dicots), such as dandelions, are much more susceptible to auxins than narrow-leaf plants (monocots) such as grasses and cereal crops, so these synthetic auxins are valuable as synthetic herbicides.

Auxins are also often used to promote initiation of adventitious roots, and are the active ingredient of the commercial preparations used in horticulture to root stem cuttings. They can also be used to promote uniform flowering and fruit set, and to prevent premature fruit drop.

Discovery of Auxin

Charles Darwin

In 1881, Charles Darwin and his son Francis performed experiments on coleoptiles, the sheaths enclosing young leaves in germinating grass seedlings. The experiment exposed the coleoptile to light from a unidirectional source and observed that they bend towards the light. By covering various parts of the coleoptiles with a light impermeable opaque cap, the Darwins discovered that light is detected by the coleoptile tip, but that bending occurs in the hypocotyl. However the seedlings showed no signs of development towards light if the tip was covered with an opaque cap, or if the tip was removed. The Darwins concluded that the tip of the coleoptile was responsible for sensing light, and proposed that a messenger is transmitted in a downward direction from the tip of the coleoptile, causing it to bend.

Peter Boysen-Jensen

In 1913 a Danish scientist named Peter Boysen-Jensen demonstrated that the signal was not transfixed but mobile. He separated the tip from the remainder of the coleoptile by a cube of gelatine which prevented cellular contact but, allowed chemicals to pass through. The seedlings responded normally bending towards the light. However when the tip was separated by an impermeable substance, there was no curvature of the stem.

Frits Went

In 1926, the Dutch botanist Frits Warmolt Went showed that a chemical messenger diffuses from coleoptile tips. Went's experiment identified how a growth promoting chemical causes a coleoptile to grow towards light. Went cut the tips of the coleoptiles and placed them in the dark, putting a few tips on agar blocks that he predicted would absorb the growth-promoting chemical. On control coleoptiles, he placed a block that lacked the chemical. On others, he placed blocks containing the chemical, either centred on top of the coleoptile to distribute the chemical evenly or offset to increase the concentration on one side. When the growth promoting chemical was distributed evenly the coleoptile grew straight. If the chemical was distributed unevenly, the coleoptile curved away from the side with the cube, as if growing towards light, even though it was grown in the dark. Went later proposed that the messenger substance is a growth-promoting hormone, which he named auxin, that becomes asymmetrically distributed in the bending region. Went concluded that auxin is at a higher concentration on the shaded side, promoting cell elongation, which results in a coleoptiles bending towards the light.

Hormonal Activity

Auxins coordinate development at all levels in plants, from the cellular level, through organs, and ultimately to the whole plant.

Molecular Mechanisms

Auxin molecules present in cells may trigger responses directly through stimulation or inhibition of the expression of sets of certain genes or by means independent of gene expression. Auxin transcriptionally activates four different families of *early genes* (aka *primary response genes*), so-called because the components required for the activation are preexisting, leading to a rapid response. The families are glutathione S-transferases, auxin homeostasis proteins like GH3, SAUR genes of currently unknown function, and the Aux/IAA repressors.

Aux/IAA, ARF, TIR1, SCF Auxin Regulatory Pathways

The Aux/IAA repressors provide an example of one of the pathways leading to auxin induced changes of gene expression. This pathway involves the protein families TIR1 (transport inhibitor response1), ARF (auxin response factor), Aux/IAA transcriptional repressors, and the ubiquitin ligase complex that is a part of the ubiquitin-proteasome protein degradation pathway. ARF proteins have DNA binding domains and can bind promoter regions of genes and activate or repress gene expression. Aux/IAA proteins can bind ARF proteins sitting on gene promoters and prevent them from doing their job. TIR1 proteins are F-box proteins that have three different domains giving them the ability to bind to three different ligands: an SCF ubiquitin ligase complex (using the F-box domain), auxin (so TIR1 proteins are auxin receptors), and Aux/IAA proteins (via a degron domain). Upon binding of auxin, a TIR1 protein's degron domain has increased affinity for Aux/IAA repressor proteins, which when bound to TIR1 and its SCF complex undergo ubiquitination and subsequent degradation by a proteasome. The degradation of Aux/IAA proteins frees ARF proteins to activate or repress genes at whose promoters they are bound.

Within a plant, elaboration of the Aux/IAA repressor pathway takes place via diversification of the TIR1, ARF, and Aux/IAA protein families. Each family may contain many similar-acting proteins, differing in qualities such as degree of affinity for partner proteins, amount of activation or repression of target gene transcription, or domains of expression (e.g. different plant tissues might express different members of the family, or different environmental stresses might activate expression of different members). Such elaboration permits the plant to use auxin in a variety of ways depending on the needs of the tissue and plant.

Other Auxin Regulatory Pathways

Another protein, auxin-binding protein 1 (ABP1), is a putative receptor for a different signaling pathway, but its role is as yet unclear. Electrophysiological experiments with protoplasts and anti-ABP1 antibodies suggest ABP1 may have a function at the plasma membrane, and cells can possibly use ABP1 proteins to respond to auxin through means faster and independent of gene expression.

On a Cellular Level

On the cellular level, auxin is essential for cell growth, affecting both cell division and cellular expansion. Auxin concentration level, together with other local factors, contributes to cell differentiation and specification of the cell fate.

Depending on the specific tissue, auxin may promote axial elongation (as in shoots), lateral expansion (as in root swelling), or isodiametric expansion (as in fruit growth). In some cases (coleoptile growth), auxin-promoted cellular expansion occurs in the absence of cell division. In other cases, auxin-promoted cell division and cell expansion may be closely sequenced within the same tissue (root initiation, fruit growth). In a living plant, auxins and other plant hormones nearly always appear to interact to determine patterns of plant development.

Organ Patterns

Growth and division of plant cells together result in growth of tissue, and specific tissue growth contributes to the development of plant organs.

Growth of cells contributes to the plant's size, unevenly localized growth produces bending, turning and directionalization of organs- for example, stems turning towards light sources (phototropism), roots growing in response to gravity (gravitropism), and other tropisms originated because cells on one side grow faster than the cells on the other side of the organ. So, precise control of auxin distribution between different cells has paramount importance to the resulting form of plant growth and organization.

Auxin Transport and the Uneven Distribution of Auxin

To cause growth in the required domains, auxins must of necessity be active preferentially in them. Auxins are not synthesized in all cells (even if cells retain the potential ability to do so, only under specific conditions will auxin synthesis be activated in them). For that purpose, auxins have to be not only translocated towards those sites where they are needed, but also they must have an established mechanism to detect those sites. For that purpose, auxins have to be translocated towards those sites where they are needed. Translocation is driven throughout the plant body, primarily from peaks of shoots to peaks of roots (from up to down).

For long distances, relocation occurs via the stream of fluid in phloem vessels, but, for short-distance transport, a unique system of coordinated polar transport directly from cell to cell is exploited. This short-distance, active transport exhibits some morphogenetic properties.

This process, polar auxin transport, is directional, very strictly regulated, and based in uneven distribution of auxin efflux carriers on the plasma membrane, which send auxins in the proper direction. Pin-formed (PIN) proteins are vital in transporting auxin.

The regulation of PIN protein localisation in a cell determines the direction of auxin transport from cell, and concentrated effort of many cells creates peaks of auxin, or auxin maxima (regions having cells with higher auxin - a maximum). Proper and timely auxin maxima within developing roots and shoots are necessary to organise the development of the organ. Surrounding auxin maxima are cells with low auxin troughs, or auxin minima. For example, in the *Arabidopsis* fruit, auxin minima have been shown to be important for its tissue development.

Organization of the Plant

As auxins contribute to organ shaping, they are also fundamentally required for proper development of the plant itself. Without hormonal regulation and organization, plants would be merely proliferating heaps of similar cells. Auxin employment begins in the embryo of the plant, where directional distribution of auxin ushers in subsequent growth and development of primary growth poles, then forms buds of future organs. Next, it helps to coordinate proper development of the arising organs, such as roots, cotyledons and leaves and mediates long distance signals between them, contributing so to the overall architecture of the plant. Throughout the plant's life, auxin helps the plant maintain the polarity of growth, and actually "recognize" where it has its branches (or any organ) connected.

An important principle of plant organization based upon auxin distribution is apical dominance, which means the auxin produced by the apical bud (or growing tip) diffuses (and is transported) downwards and inhibits the development of ulterior lateral bud growth, which would otherwise compete with the apical tip for light and nutrients. Removing the apical tip and its suppressively acting auxin allows the lower dormant lateral buds to develop, and the buds between the leaf stalk and stem produce new shoots which compete to become the lead growth. The process is actually quite complex, because auxin transported downwards from the lead shoot tip has to interact with several other plant hormones (such as strigolactones or cytokinins) in the process on various positions along the growth axis in plant body to achieve this phenomenon. This plant behaviour is used in pruning by horticulturists.

Finally, the sum of auxin arriving from stems to roots influences the degree of root growth. If shoot tips are removed, the plant does not react just by outgrowth of lateral buds — which are supposed to replace to original lead. It also follows that smaller amount of auxin arriving to the roots results in slower growth of roots and the nutrients are subsequently in higher degree invested in the upper part of the plant, which hence starts to grow faster.

The Five Models for Auxin Transport

Five models of auxin transport have been made, using *Arabidopsis* as the study plant.

First Model

In the first model incoming light deactivates auxin on the light side of the plant allowing the shaded part to continue growing and eventually bend the plant over towards the light.

Second Model

In the second model light inhibits auxin biosynthesis on the light side of the plant, thus decreasing the concentration of auxin relative to the unaffected side.

Third Model

In the third model there is a horizontal flow of auxin from both the light and dark side of the plant. Incoming light causes more auxin to flow from the exposed side to the shaded side, increasing the concentration of auxin on the shaded side and thus more growth occurring.

Fourth Model

In the fourth model shows receiving light to inhibit auxin basipetal down to the exposed side, causing the auxin to only flow down the shaded side.

Fifth Model

Model five encompasses elements of both model 3 and 4. The main auxin flow in this model comes from the top of the plant vertically down towards the base of the plant with some of the auxin travelling horizontally from the main auxin flow to both sides of the plant. Receiving light inhibits the horizontal auxin flow from the main vertical auxin flow to the irradiated exposed side.

Effects

Auxin participates in phototropism, geotropism, hydrotropism and other developmental changes. The uneven distribution of auxin, due to environmental cues, such as unidirectional light or gravity force, results in uneven plant tissue growth, and generally, auxin governs the form and shape of plant body, direction and strength of growth of all organs, and their mutual interaction.

Figure: *A healthy* Arabidopsis thaliana *plant (left) next to an auxin signal-transduction mutant*

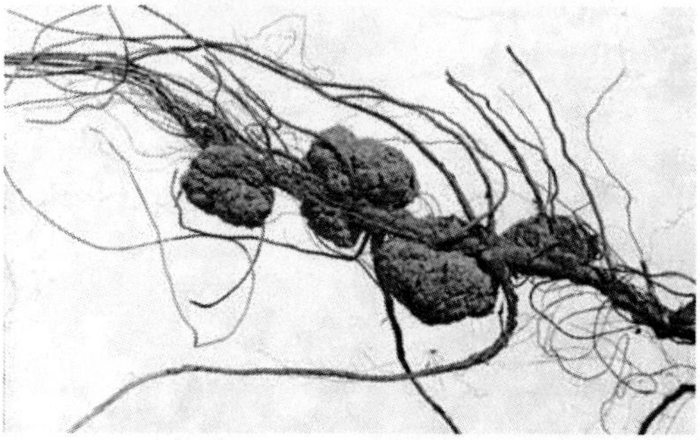

Figure: *Crown galls are caused by* Agrobacterium tumefaciens *bacteria; they produce and excrete auxin and cytokinin, which interfere with normal cell division and cause tumors.*

Auxin stimulates cell elongation by stimulating wall-loosening factors, such as elastins, to loosen cell walls. The effect is stronger if gibberellins are also present. Auxin also stimulates cell division if cytokinins are present. When auxin and cytokinin are applied to callus, rooting can be generated if the auxin concentration is higher than cytokinin concentration. Xylem tissues can be generated when the auxin concentration is equal to the cytokinins.

Auxin also induces sugar and mineral accumulation at the site of application.

Wound Response

Auxin induces the formation and organization of phloem and xylem. When the plant is wounded, the auxin may induce the cell differentiation and regeneration of the vascular tissues.

Root Growth and Development

Auxins promote root initiation. Auxin induces both growth of pre-existing roots and adventitious root formation, i.e., branching of the roots. As more native auxin is transported down the stem to the roots, the overall development of the roots is stimulated. If the source of auxin is removed, for example the tips of stems are trimmed, the roots are less stimulated accordingly, and growth of stem is supported instead.

In horticulture, auxins, especially NAA and IBA, are commonly applied to stimulate root initiation when rooting cuttings of plants. However, high concentrations of auxin inhibit root elongation and instead enhance adventitious root formation. Removal of the root tip can lead to inhibition of secondary root formation.

Apical Dominance

Auxin induces shoot apical dominance; the axillary buds are inhibited by auxin, as a high concentration of auxin directly stimulates ethylene synthesis in lateral buds, causing inhibition of their growth and potentiation of apical dominance. When the apex of the plant is removed, the inhibitory effect is removed and the growth of lateral buds is enhanced. Auxin is sent to the part of the plant facing light, and this promotes growth towards that direction.

Fruit Growth and Development

Auxin is required for fruit growth and development and delays fruit senescence. When seeds are removed from strawberries, fruit growth is stopped; exogenous auxin stimulates the growth in fruits with seeds removed. For fruit with unfertilized seeds, exogenous auxin results in parthenocarpy ("virgin-fruit" growth).

Fruits form abnormal morphologies when auxin transport is disturbed. In *Arabidopsis* fruits, auxin controls the release of seeds from the fruit (pod). The valve margins are a specialised tissue in pods that regulates when pod will open (dehiscence). Auxin must be removed from the valve margin cells to allow the valve margins to form. This process requires modification of the auxin transporters (PIN proteins).

Flowering

Auxin plays also a minor role in the initiation of flowering and development of reproductive organs. In low concentrations, it can delay the senescence of flowers.

A number of plant mutants have been described that affect flowering and have deficiencies in either auxin synthesis or transport. In maize, one example is bif2 barren inflorescence2.

Ethylene Biosynthesis

In low concentrations, auxin can inhibit ethylene formation and transport of precursor in plants; however, high concentrations can induce the synthesis of ethylene. Therefore, the high concentration can induce femaleness of flowers in some species.

Auxin inhibits abscission prior to formation of abscission layer, and thus inhibits senescence of leaves.

Synthetic Auxins

In the course of research on auxin biology, many compounds with noticeable auxin activity were synthesized. Many of them had been found to have economical potential for man-controlled growth and development of plants in agronomy. Synthetic auxins include the following compounds:

Auxins are toxic to plants in large concentrations; they are most toxic to dicots and less so to monocots. Because of this property, synthetic auxin herbicides, including 2,4-D and 2,4,5-T, have been developed and used for weed control.

However, synthetic auxins, especially 1-naphthaleneacetic acid (NAA) and indole-3-butyric acid (IBA), are also commonly applied to stimulate root growth when taking cuttings of plants or for different agricultural purposes such as the prevention of fruit drop in orchards.

Used in high doses, auxin stimulates the production of ethylene. Excess ethylene (also native plant hormone) can inhibit elongation growth, cause leaves to fall (abscission), and even kill the plant. Some synthetic auxins, such as 2,4-D and 2,4,5-trichlorophenoxyacetic acid (2,4,5-T) were marketed also as herbicides. Dicots, such as dandelions, are much more susceptible to auxins than monocots), such as grasses and cereal crops. So, these synthetic auxins are valuable as synthetic herbicides. 2,4-D was the first widely used herbicide, and it is still so.

2,4-D was first commercialized by the Sherwin-Williams company, and saw use in the late 1940s. It is easy and inexpensive to manufacture.

Herbicide Manufacture

The defoliant Agent Orange, used extensively by British forces in the Malayan Emergency and American forces in the Vietnam War, was a mix of 2,4-D and 2,4,5-T.

The compound 2,4-D is still in use and is thought to be safe, but 2,4,5-T was more or less banned by the U.S. Environmental Protection Agency in 1979. The dioxin TCDD is an unavoidable contaminant produced in the manufacture of 2,4,5-T.

As a result of the integral dioxin contamination, 2,4,5-T has been implicated in leukemia, miscarriages, birth defects, liver damage, and other diseases.

Bibliography

Anon (Inderjit): *Allelopathy, from Molecules to Ecosystems*, Science Publishers, Enfield, New Hampshire, 2002.

Balfour, Lady Eve B. *The Living Soil,* Faber and Faber, London, 1943.

Bennett, Hugh Hammond : *Elements of Soil Conservation*, Biotech Books, Delhi, 2009.

Bolton, Malcolm D.: *A Guide to Soil Mechanics*, Universities Press, Delhi, 2003.

Chandra, Ramesh and Satish Kumar Singh: *Fundamentals and Management of Soil Quality*, Westville Pub, Delhi, 2009.

Dar, Ghulam Hassan: *Soil Microbiology and Biochemistry*, New India Publishing Agency, Delhi, 2010.

Das, Braja M. : *Advanced Soil Mechanics*, Taylor and Francis, Delhi, 2010.

Einhellig, F.A.: *Allelopathy, from Molecules to Ecosystems*, Science Publishers, Enfield, New Hampshire, 2002.

Foth, Henry D. *Fundamentals of Soil Science,* Eighth Edition. New York: John Wylie & Sons, 1990.

Gupta, P.K. : *A Handbook of Soil Fertilizer and Manure*, Agrobios, Delhi, 2011.

Hopkins, Cyril G. *Soil Fertility and Permanent Agriculture.* Boston: Ginn and Company, 1910.

Jose S.: *Chemical Ecology of Plants: Allelopathy in Aquatic and Terrestrial Ecosystems*, Birkhauser Verlag, Basel, Switzerland, 2002.

Macself, R : *Soils and Fertilizers*, Satish Serial Pub, Delhi, 2005.

Majumdar, S.P. and R.A. Singh: *Analysis of Soil Physical Properties*, Agrobios, Delhi, 2008.

Malcolm D. Bolton: *A Guide to Soil Mechanics*, Universities Press, Delhi, 2003.

Mallik, A. U. and Anon: *Chemical Ecology of Plants: Allelopathy in Aquatic and Terrestrial Ecosystems*, Birkhauser Verlag, Basel, Switzerland, 2002.

Mathur, S K B B S Kapoor: *Emerging Trends In Soil Management*, Madhu Publications, Delhi, 2007.

Narwal, S.S. : *Allelopathy In Soil Sickness*, Scientific, Delhi, 2006.

Patel, Sheelwant: *Indian Forests : Soil Water and Bio-Environment Conservation*, Pointer, Delhi, 2005.

Rao, A.V. Narasimha: *Fundamentals of Soil Mechanics*, Laxmi Publications, Delhi, 2010.

Reigosa, M. J., N. Pedrol, A. M. Sanchez-Moreiras, and L. Gonzales: *Allelopathy, from Molecules to Ecosystems*, Science Publishers, Enfield, New Hampshire, 2002.

Russell, E.J. : *Handbook on Soils and Manures*, Discovery, 1999.

Singh, Lalit: *Polymer Microscopy*, London: Hurst and Company Publishers, 2001.

Somani, L L and P C Kanthaliya: *Soils and Fertilisers at a Glance*, Agrotech, Delhi, 2004.

Stone, D. C.: *Instrumental Analysis Laboratory Manual*, Department of Chemistry, University of Toronto, 2005.

Trevor H. : *Transforming Matter: A History of Chemistry from Alchemy to the Buckyball,* Baltimore: Johns Hopkins University Press, 2001.

Vasudev Bhatnagar: *Cell Science and Technology*, Campus Books, Delhi, 2009.

Walters, P.: *The New Quantum Universe*, University Press, Cambridge, 2004.

Willis, R. J.: *Principles and Practices in Plant Ecology: Allelochemical Interactions*, CRC Press, and Boca Raton, FL, 1999.

Index

H

I

L

M

N

O

P

R

S

T

V

W

❑❑❑